Sandra Tisdell–Clifford

DEVELOPMENTAL MATHEMATICS
BOOK 4

Founding authors
Allan Thompson · Effie Wrightson

Series editor
Robert Yen

NELSON
A Cengage Company

Australia • Brazil • Japan • Korea • Mexico • Singapore • Spain • United Kingdom • United States

Developmental Mathematics Book 4
5th Edition
Sandra Tisdell-Clifford

Publisher: Robert Yen
Project editor: Alan Stewart
Editor: Anna Pang
Text design: Sarah Hazell
Cover design: Sarah Hazell
Permissions researcher: Flora Smith
Production controller: Erin Dowling
Cover image: iStockphoto/shuoshu; shutterstock.com/javarman
Typeset by: Cenveo Publishing Services

Any URLs contained in this publication were checked for currency during the production process. Note, however, that the publisher cannot vouch for the ongoing currency of URLs.

© 2015 Cengage Learning Australia Pty Limited

For product information and technology assistance,
in Australia call **1300 790 853**;
in New Zealand call **0800 449 725**

For permission to use material from this text or product, please email
aust.permissions@cengage.com

National Library of Australia Cataloguing-in-Publication Data
Tisdell-Clifford, Sandra, author.
Developmental mathematics. Book 4 / Sandra Tisdell-Clifford.

5th edition.
9780170351058 (paperback)
For secondary school age.

Mathematics-Study and teaching (Secondary)--Australia.
Mathematics--Problems, exercises, etc

510.76

Cengage Learning Australia
Level 7, 80 Dorcas Street
South Melbourne, Victoria Australia 3205

Cengage Learning New Zealand
Unit 4B Rosedale Office Park
331 Rosedale Road, Albany, North Shore 0632, NZ

For learning solutions, visit **cengage.com.au**

Printed in China by 1010 Printing International Limited
7 8 24

CONTENTS

* = NSW STAGE 5.2, # = NSW additional content

* = NSW STAGE 5.2, # = NSW additional content

* = NSW STAGE 5.2, # = NSW additional content

PREFACE

In schools for over four decades, *Developmental Mathematics* is a unique, well-known and trusted Years 7–10 mathematics series with a strong focus on key numeracy and literacy skills. This 5th edition of the series has been revised for the new Australian curriculum as well as the NSW syllabus Stages 4 and 5.1. The four books of the series contain short chapters with worked examples, definitions of key words, graded exercises, a language activity and a practice test. Each chapter covers a topic that should require about two weeks of teaching time.

Developmental Mathematics supports students with mathematics learning, encouraging them to experience more confidence and success in the subject. This series presents examples and exercises in clear and concise language to help students master the basics and improve their understanding. We have endeavoured to equip students with the essential knowledge required for success in junior high school mathematics, with a focus on basic skills and numeracy.

Developmental Mathematics Book 4 is written for students in Year 10, covering the Australian curriculum and NSW syllabus (see the curriculum grids on the following pages and the teaching program on the *NelsonNet* teacher website). This book presents concise and highly-structured examples and exercises, with each new concept or skill on a double-page spread for convenient reading and referencing.

Students learning mathematics need to be taught by dynamic teachers who use a variety of resources. Our intention is that teachers and students use this book as their primary source or handbook, and supplement it with additional worksheets and resources, including those found on the *NelsonNet* teacher website (access conditions apply). We hope that teachers can use this book effectively to help students achieve success in secondary mathematics. Good luck!

ABOUT THE AUTHOR

Sandra Tisdell-Clifford teaches at Newcastle Grammar School and was the Mathematics coordinator at Our Lady of Mercy College (OLMC) in Parramatta for 10 years. Sandra is best known for updating *Developmental Mathematics* for the 21st century (4th edition, 2003) and writing its blackline masters books. She also co-wrote *Nelson Senior Maths 11 General* for the Australian curriculum, teaching resources for the NSW senior series *Maths In Focus* and the Years 7–8 homework sheets for *New Century Maths / NelsonNet*.

Sandra expresses her thanks and appreciation to the Headmaster and staff of Newcastle Grammar School and dedicates this book to her husband, Ray Clifford, for his support and encouragement. She also thanks series editor **Robert Yen** and editors **Anna Pang** and **Alan Stewart** at Cengage Learning for their leadership on this project.

Original authors **Allan Thompson** and **Effie Wrightson** wrote the first three editions of *Developmental Mathematics* (1974, 1981 and 1988) and taught at Smith's Hill High School in Wollongong. Sandra thanks them for their innovative pioneering work, which has paved the way for this new edition for the Australian curriculum.

ISBN 9780170351058

FEATURES OF THIS BOOK

- Each chapter begins with a table of contents and list of chapter outcomes
- Each teaching section of a chapter is presented clearly on a double-page spread

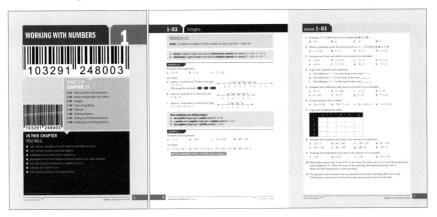

- The left page contains explanations, worked examples, and if appropriate, a Wordbank of mathematical terminology and a fact box
- The right page contains an exercise set, including multiple-choice questions, scaffolded solutions and realistic applications of mathematics
- Each chapter concludes with a **Language activity** (puzzle) that reinforces mathematical terminology in a fun way, and a **Practice test** containing non-calculator questions on general topics and topic questions grouped by chapter subheading

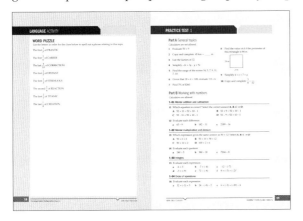

- **Answers** and **index** are at the back of the book
- **NelsonNet teacher website www.nelsonnet.com.au** of additional worksheets, puzzle sheets, skillsheets, video tutorials, technology worksheets, teaching program, curriculum grids, chapter PDFs of this book

- Note: NelsonNet access is available to teachers who use Developmental Mathematics as a core educational resource in their classroom. Contact your sales representative for information about access codes and conditions.

CURRICULUM GRID
AUSTRALIAN CURRICULUM

STRAND AND SUBSTRAND	DEVELOPMENTAL MATHEMATICS BOOK 1 CHAPTER		DEVELOPMENTAL MATHEMATICS BOOK 2 CHAPTER	
NUMBER AND ALGEBRA				
Number and place value	1	Integers and the number plane	1	Working with numbers
	3	Working with numbers	2	Primes and powers
	4	Factors and primes	4	Integers
	5	Powers and decimals		
	6	Multiplying and dividing decimals		
	9	Algebra and equations		
Real numbers	5	Powers and decimals	5	Decimals
	6	Multiplying and dividing decimals	11	Fractions
	7	Fractions	12	Percentages
	8	Multiplying and dividing fractions	16	Ratios and rates
	17	Percentages and ratios		
Money and financial mathematic	6	Multiplying and dividing decimals	12	Percentages
Patterns and algebra	9	Algebra and equations	6	Algebra
			15	Further algebra
Linear and non-linear relationships	1	Integers and the number plane	15	Further algebra
	9	Algebra and equations	17	Graphing lines
MEASUREMENT AND GEOMETRY				
Using units of measurement	12	Length and time	9	Length and time
	13	Area and volume	10	Area and volume
Shape	10	Shapes and symmetry		
Location and transformation	10	Shapes and symmetry	7	Angles and symmetry
Geometric reasoning	2	Angles	7	Angles and symmetry
	11	Geometry	8	Triangles and quadrilaterals
Pythagoras and trigonometry			3	Pythagoras' theorem
STATISTICS AND PROBABILITY				
Chance	16	Probability	14	Probability
Data representation and interpretation	14	Statistical graphs	14	Investigating data (AC)
	15	Analysing data		

ISBN 9780170361050

CURRICULUM GRID
AUSTRALIAN CURRICULUM

STRAND AND SUBSTRAND	DEVELOPMENTAL MATHEMATICS BOOK 3 CHAPTER		DEVELOPMENTAL MATHEMATICS BOOK 4 CHAPTER	
NUMBER AND ALGEBRA				
Real numbers	2	Whole numbers and decimals	1	Working with numbers
	3	Integers and fractions	2	Percentages
	6	Percentages	7	Ratios and rates
	7	Indices		
	16	Ratios and rates		
Money and financial mathematics	6	Percentages	3	Earning and saving money
Patterns and algebra	4	Algebra	4	Algebra
	7	Indices	10	Indices
Linear and non-linear relationships	9	Equations	13	Equations and inequalities
	14	Graphing lines	15	Coordinate geometry
			16	Graphing lines and curves
MEASUREMENT AND GEOMETRY				
Using units of measurement	12	Length and time	9	Length and time
	13	Area and volume	11	Area and volume
Geometric reasoning	8	Geometry	8	Congruent and similar figures
Pythagoras and trigonometry	1	Pythagoras' theorem	5	Pythagoras' theorem
	5	Trigonometry	6	Trigonometry
STATISTICS AND PROBABILITY				
Chance	15	Probability	14	Probability
Data representation and interpretation	11	Investigating data	12	Investigating data

Note: These more abstract concepts are not covered in the *Developmental Mathematics* series:

- Binomial products and factors
- Algebraic fractions
- Simultaneous equations
- Quadratic equations
- Equations of parallel and perpendicular lines
- Direct proportion
- Two-step chance experiments and tree diagrams
- Conditional probability
- Scatter plots
- Sample vs population means

SERIES OVERVIEW

ISBN 9780170351058

WORKING WITH NUMBERS

103291 248003

103291 248003

IN THIS CHAPTER YOU WILL:

- add, subtract, multiply and divide mentally with whole numbers
- add, subtract, multiply and divide integers
- understand and use the order of operations
- understand and convert between improper fractions and mixed numerals
- calculate equivalent fractions and simplify fractions
- compare and order fractions
- add, subtract, multiply and divide fractions

Shutterstock.com/Nickylarson974

EXAMPLE 1

Find the value of each expression.

a $23 + 58$ **b** $86 - 48$

SOLUTION

a $23 + 58 = 20 + 3 + 50 + 8$ or

$$= 20 + 50 + 3 + 8$$
$$= 70 + 11$$
$$= 81$$

$$2^{1}3 +$$
$$\underline{58}$$
$$\underline{81}$$

b $86 - 48 = 80 + 6 - 40 - 8$ or

$$= 80 - 40 + 6 - 8$$
$$= 40 - 2$$
$$= 38$$

$$^{7}8^{1}6 -$$
$$\underline{4\ 8}$$
$$\underline{3\ 8}$$

Mental addition and subtraction strategies

▨ Look for units digits that add up to 10, such as 3 and 7.

▨ If one of the numbers is close to 10, 20, 30, ..., split it up.

EXAMPLE 2

Evaluate each sum mentally by pairing up numbers that have units digits adding to 10.

a $33 + 66 + 17 + 4$ **b** $224 + 382 + 476$

SOLUTION

a $33 + 66 + 17 + 4 = (33 + 17) + (66 + 4)$

$$= 50 + 70$$
$$= 120$$

b $224 + 382 + 476 = (224 + 476) + 382$

$$= 700 + 382$$
$$= 1082$$

EXAMPLE 3

Evaluate each expression mentally by splitting up the second number.

a $78 + 11$ **b** $72 + 99$ **c** $94 - 9$ **d** $235 - 51$

SOLUTION

a $78 + 11 = 78 + 10 + 1$

✱ To add 11, add 10 and then 1.

$$= 88 + 1$$
$$= 89$$

b $72 + 99 = 72 + 100 - 1$

✱ To add 99, add 100 and subtract 1.

$$= 172 - 1$$
$$= 171$$

c $94 - 9 = 94 - 10 + 1$

✱ To subtract 9, subtract 10 and add 1.

$$= 84 + 1$$
$$= 85$$

d $235 - 51 = 235 - 50 - 1$

✱ To subtract 51, subtract 50 and then 1.

$$= 185 - 1$$
$$= 184$$

ISBN 9780170351058

1   Which expression gives the same answer as 82 + 15? Select the correct answer **A**, **B**, **C** or **D**.
   A 82 + 51   **B** 80 + 2 + 15   **C** 28 + 15   **D** 82 + 10 + 3

2   Which expression gives the same answer as 65 – 12? Select **A**, **B**, **C** or **D**.
   A 65 – 10 – 2   **B** 65 – 10 + 2   **C** 60 + 5 – 10 + 2   **D** 65 + 10 – 2

3   Evaluate each sum using mental calculation.
   a 14 + 7   **b** 23 + 15   **c** 38 + 5   **d** 24 + 128
   e 18 + 22   **f** 32 + 27   **g** 46 + 16   **h** 78 + 19

4   Which expression gives the same answer as 375 + 499? Select **A**, **B**, **C** or **D**.
   A 357 + 499   **B** 375 + 949   **C** 375 + 500 – 1   **D** 375 + 480 + 9

5   Evaluate each expression without the use of a calculator.
   a 27 + 64   **b** 64 + 27   **c** 27 + 64 + 36   **d** 64 + 27 + 36 + 48

6   **a**   How could you use your answer to **5 a** to find **5 b**?
   b   How could you use your answer to **5 c** to find **5 d**?

7   Evaluate each difference using mental calculation.
   a 19 – 4   **b** 38 – 12   **c** 65 – 24   **d** 84 – 28
   e 52 – 48   **f** 86 – 24   **g** 174 – 98   **h** 296 – 67

8   Evaluate each expression by pairing up numbers that have units digits adding to 10.
   a 22 + 63 + 8 + 27   **b** 84 + 45 – 5 + 6
   c 48 + 52 + 32 – 2   **d** 96 + 35 – 15 + 5
   e 243 – 23 + 17   **f** 421 + 512 + 9
   g 381 + 289 – 19   **h** 503 – 23 + 112

9   Copy and complete the following.
   a   75 + 11 = 75 + ___ + 1   **b**   278 – 41 = 278 – ___ – 1
              = 85 + ___   = 238 – ___
              = _____   = _____

10 Evaluate each expression mentally by splitting up the second number.
   a 52 – 9   **b** 68 + 11   **c** 84 – 21   **d** 54 + 9
   e 122 – 31   **f** 186 + 19   **g** 142 – 19   **h** 165 + 29

11 Is each statement true or false?
   a 64 – 29 = 35   **b** 75 – 41 = 24   **c** 135 – 82 = 43   **d** 286 – 94 = 192

WORDBANK

product The answer to a multiplication (×) of two or more numbers.

quotient The answer to a division (÷) of two numbers.

Mental multiplication strategies

Multiplying by	Strategy
2	Double
4	Double twice
5	Multiply by 10, then halve
8	Multiply by (10 – 2) or double three times
9	Multiply by (10 – 1)
10	Add 0 to the end
11	Multiply by (10 + 1)
12	Multiply by (10 + 2)

Mental division strategies

Dividing by	Strategy
2	Halve
4	Halve twice
5	Divide by 10, then double
8	Halve 3 times
10	Move the decimal point 1 place left, or for a whole number ending in 0, drop a 0 from the end of the number
20	Divide by 10, then halve
100	Move the decimal point 2 places left, or for a whole number ending in 0s, drop two 0s from the end of the number

EXAMPLE 4

Evaluate each product mentally.

a 28×5 **b** 43×12 **c** $54 \times 50 \times 2$

SOLUTION

a $28 \times 5 = 28 \times 10 \div 2$

✱ because 5 is $\frac{1}{2}$ of 10

$= 280 \div 2$
$= 140$

b $43 \times 12 = 43 \times (10 + 2)$
$= 43 \times 10 + 43 \times 2$
$= 430 + 86$
$= 516$

c $54 \times 50 \times 2 = 54 \times (50 \times 2)$

✱ numbers can be multiplied in any order

$= 54 \times 100$
$= 5400$

EXAMPLE 5

Evaluate each quotient mentally.

a $660 \div 5$ **b** $1600 \div 20$ **c** $872 \div 8$

SOLUTION

a $660 \div 5 = 660 \div 10 \times 2$

✱ divide by 10 and double

$= 66 \times 2$
$= 132$

b $1600 \div 20 = 1600 \div 10 \div 2$

✱ divide by 10, then halve

$= 160 \div 2$
$= 80$

c $872 \div 8 = 872 \div 2 \div 2 \div 2$

✱ halve 872 three times

$= 436 \div 2 \div 2$
$= 218 \div 2$
$= 109$

1   Which expression can be used to evaluate 63×8 mentally? Select the correct answer **A**, **B**, **C** or **D**.

   A $63 \times (10 - 1)$ **B** $63 \times (10 - 2)$

   C $63 \times (10 + 1)$ **D** $63 \times (10 + 2)$

2   Which mental method can be used to divide a number by 20? Select **A**, **B**, **C** or **D**.

   A halve it 10 times   **B** divide by 10 then double

   C divide by 10 twice   **D** divide by 10 then halve

3   Evaluate each expression.

a 4×6	**b** 8×5	**c** 7×3	**d** 8×2
e 5×9	**f** 4×6	**g** 9×8	**h** 4×7
i 6×8	**j** 10×4	**k** 6×7	**l** 8×6
m $18 \div 3$	**n** $30 \div 6$	**o** $54 \div 9$	**p** $25 \div 5$
q $49 \div 7$	**r** $56 \div 8$	**s** $35 \div 7$	**t** $28 \div 4$
u $33 \div 3$	**v** $80 \div 10$	**w** $63 \div 9$	**x** $64 \div 8$

4   Evaluate each product using a mental strategy.

a 14×5	**b** 36×11	**c** 58×9	**d** 23×4
e 45×100	**f** 64×8	**g** 93×2	**h** 74×12
i 16×20	**j** 43×10	**k** 52×40	**l** 31×9
m 56×5	**n** 42×4	**o** 25×11	**p** 72×4

5   Evaluate each quotient using a mental strategy.

a $58 \div 2$	**b** $136 \div 4$	**c** $168 \div 8$	**d** $524 \div 4$
e $184 \div 8$	**f** $218 \div 2$	**g** $90 \div 5$	**h** $320 \div 10$
i $800 \div 5$	**j** $490 \div 10$	**k** $950 \div 5$	**l** $760 \div 20$
m $1600 \div 100$	**n** $280 \div 8$	**o** $55 \div 10$	**p** $2380 \div 100$
q $254 \div 10$	**r** $32\,600 \div 100$	**s** $9800 \div 10$	**t** $6.5 \div 100$

6   Evaluate each product by changing the order.

   a $2 \times 29 \times 5$ **b** $25 \times 18 \times 4$

   c $2 \times 46 \times 50$ **d** $5 \times 62 \times 20$

   e $4 \times 9 \times 5$ **f** $5 \times 21 \times 200$

   g $2 \times 15 \times 6$ **h** $25 \times 13 \times 4$

WORDBANK

integer A positive or negative whole number, or zero, such as 8, –1 and –10.

- **Adding** a negative integer is the same as **subtracting its opposite**; for example, 3 + (–4) = 3 – 4 = –1
- **Subtracting** a negative integer is the same as **adding its opposite**; for example, 3 – (–4) = 3 + 4 = 7

EXAMPLE 6

Evaluate each expression.

a $-3 + 5$ **b** $1 - 4$ **c** $-2 - (-3)$

SOLUTION

a Start at –3 and move 5 units to the right
$-3 + 5 = 2$
OR using the calculator:

b Start at 1 and move 4 units to the left
$1 - 4 = -3$

c Start at –2 and move 3 units to the right.
$-2 - (-3) = -2 + 3 = 1$

When multiplying and dividing integers:
- **two positive** integers give a **positive** answer (**+ + = +**)
- a **positive** and a **negative** integer give a **negative** answer (**+ – = –**)
- **two negative** integers give a **positive** answer (**– – = +**)

EXAMPLE 7

Evaluate each expression.

a -7×6 **b** $(-9)^2$ **c** $72 \div (-6)$ **d** $-48 \div (-8)$

SOLUTION

a $-7 \times 6 = -42$

b $(-9)^2 = -9 \times (-9) = 81$

c $72 \div (-6) = -12$

d $-48 \div (-8) = 6$

✳ Check that these answers are correct on your calculator.

1   Evaluate –9 + 5. Select the correct answer **A**, **B**, **C** or **D**.

    A –14   **B** 4   **C** –4   **D** 9

2   Which expression gives the same answer as –6 – (–2)? Select **A**, **B**, **C** or **D**.

    A –6 + (–2)   **B** 6 + 2   **C** –6 + 2   **D** 6 + (–2)

3   Evaluate each sum and check your answer on a calculator.

    a –2 + 8   **b** 3 + (–5)   **c** –3 + 7   **d** 9 + (–6)

    e –4 + 4   **f** 7 + (–9)   **g** –8 + 3   **h** –7 + 11

4   Copy and complete each statement.

    a The difference 7 – (–3) is the same as the sum 7 + ___.

    b The difference 7 – 3 is the same as the sum _____.

    c The difference –7 – 3 is the same as the sum _____.

5   Evaluate each difference and check your answer on a calculator.

    a 12 – 5   **b** 2 – (–4)   **c** –3 – 7   **d** 5 – 8

    e 4 – (–6)   **f** –8 – 3   **g** 3 – 11   **h** 9 – (–5)

6   Is each equation true or false?

    a –8 × 3 = 24   **b** –24 ÷ 3 = –8

    c 24 ÷ (–8) = 3   **d** –3 × 8 = 24

7   Copy and complete this table.

×	-3	7	-5	9	-4
2					
-8					
-6					
11					

8   Evaluate each quotient and check your answer on a calculator.

    a –27 ÷ 3   **b** 64 ÷ (–8)   **c** –28 ÷ (–7)   **d** 42 ÷ (–6)

    e –81 ÷ (–9)   **f** 40 ÷ (–8)   **g** –48 ÷ 8   **h** –55 ÷ 5

9   Evaluate each expression and check your answer on a calculator.

    a $(-3)^2$ **b** $(-3)^3$ **c** $(-3)^4$ **d** $-4 \times (-5)^2$

10 When Jake went to bed, it was 19°C in his room. He woke up at 3 a.m. and the temperature had dropped by 4°. When he woke in the morning, the temperature had risen 7°. What was the temperature in the morning?

11 Tia opened a new business and was shocked to find that she was losing $420 each week. If this trend continued for the first 8 weeks, how much money had she lost?

WORDBANK

brackets Grouping symbols around expressions such as round brackets ( ) or square brackets [ ].

index or power A small number placed to the top right of a number. For example, $3^2 = 3 \times 3$. Here, 3 is the base number and 2 is the index or power.

square root The opposite of squaring a number. For example, $\sqrt{16} = 4$ as $4^2 = 4 \times 4 = 16$.

When evaluating **mixed expressions** with more than one operation, calculate using this **order of operations**:
- ▨ brackets () first
- ▨ then powers (x^y) and square roots ($\sqrt{\ }$)
- ▨ then multiplication (×) and division (÷) from left to right
- ▨ then addition (+) and subtraction (−) from left to right.

EXAMPLE 8

Evaluate each mixed expression using the order of operations.

a $40 - 24 \div (-3)$

b $58 + (-6) \times (-8)$

c $(-7 + 2) \times (-54 \div 9)$

d $7.5 \div 5 + (-4) \times 0.9$

e $(-6)^2 \div (-9) + \sqrt{25} \times (-3)$

f $\dfrac{-56 + 24}{76 - 68}$

SOLUTION

a $\begin{aligned} 40 - 24 \div (-3) &= 40 - (-8) \\ &= 48 \end{aligned}$ Division first

b $\begin{aligned} 58 + (-6) \times (-8) &= 58 + 48 \\ &= 106 \end{aligned}$ Multiplication first

c $\begin{aligned} (-7 + 2) \times (-54 \div 9) &= -5 \times (-6) \\ &= 30 \end{aligned}$ Brackets first

d $\begin{aligned} 7.5 \div 5 + (-4) \times 0.9 &= 1.5 + (-3.6) \\ &= -2.1 \end{aligned}$ Divide and multiply first

e $\begin{aligned} (-6)^2 \div (-9) + \sqrt{25} \times (-3) &= 36 \div (-9) + 5 \times (-3) \\ &= -4 + (-15) \\ &= -19 \end{aligned}$ Powers and $\sqrt{\ }$ first

f $\begin{aligned} \dfrac{-56 + 24}{76 - 68} &= \dfrac{-32}{8} \\ &= -4 \end{aligned}$ Work out the top and bottom first

1   In a mixed expression, which operation comes first? Select the correct answer **A**, **B**, **C** or **D**.

   A multiplication   **B** addition   **C** brackets   **D** powers

2   For $(24 - 8) \times (-3) + 7^2$, which operation is performed first? Select **A**, **B**, **C** or **D**.

   A $24 - 8$ **B** $8 \times (-3)$ **C** $(-3) + 7^2$ **D** 7^2

3   Which operation is done first in each expression below?

   a $20 - 6 \times (-3)$ **b** $54 + (-6) \div (-2)$ **c** $72 - (-49) \div 7$

   d $56 + 42 \div (-6)$ **e** $-32 + (-7) \times 4$ **f** $90 - 65 \div (-5)$

4   Evaluate each expression in Question **3**.

5   Evaluate each expression.

   a $(-8 \times 2) + [24 \div (-3)]$ **b** $-55 + (-9 + 4) \times (-5)$ **c** $280 - (-6 + 12) \div (-3)$

   d $84 - (-8) \times (-2 + 6)$ **e** $(-8 \times -7) \div (-2 - 2)$ **f** $[-42 \div (-7)] \times (-4 + 9)$

6   Evaluate each expression using your calculator.

   a $\{580 - [-22 + (-28)]\} \div 70$ **b** $-322 \div [-12 + (-8 \times 5) + 6]$

   c $120 - [-48 \times (-3 + 5)] + (-8)$ **d** $[270 + (-15 + 35)] - 56$

7   Evaluate each expression.

   a $(-6)^2 \times (-4) + 12$ **b** $-72 \div (-9) \times 4.2$ **c** $24.8 - (-3)^2 + 6.4$

   d $32.8 - (-4) \times 12.2$ **e** $\sqrt{16} \times (-7) + (-2)^2$ **f** $46.8 - 14.21 \div (-7)$

   g $\sqrt{64} + (-6)^2 \times (-2)$ **h** $\dfrac{86 - 18}{28 \div 7}$ **i** $\dfrac{26.4 - 12.6}{36 \div (-12)}$

8   A sailing club charges $180 per year for membership, $26.50 to sail on weekdays and $34 to sail on the weekend. How much will it cost Ray over a year if he sails twice during the week and once every second weekend?

iStockphoto.com/photovideostock

EXAMPLE 9

a Change $\dfrac{7}{4}$ to a mixed numeral.

b Change $1\dfrac{2}{3}$ to an improper fraction.

SOLUTION

a $\dfrac{7}{4} = 7 \div 4$

$= 1$ remainder 3

$= 1\dfrac{3}{4}$

✱ Write the remainder in the numerator of the fraction.

OR on a calculator, enter: 7 (a^b/c) 4 (=)

b $1\dfrac{2}{3} = \dfrac{1 \times 3}{3} + \dfrac{2}{3}$

✱ change 1 to a fraction: $1 = \dfrac{3}{3}$

$= \dfrac{3}{3} + \dfrac{2}{3}$

$= \dfrac{5}{3}$

OR on a calculator, enter: 1 (a^b/c) 2 (a^b/c) 3 (=) (d/c) *

*The improper fraction key (d/c) or (□□/□) may require the (SHIFT) or (2ndF) key.

To find equivalent fractions, multiply the numerator and denominator by the same number.

EXAMPLE 10

Form 3 equivalent fractions for $\dfrac{2}{5}$.

SOLUTION

$\dfrac{2}{5} = \dfrac{2 \times 2}{5 \times 2} = \dfrac{4}{10}$ $\dfrac{2}{5} = \dfrac{2 \times 3}{5 \times 3} = \dfrac{6}{15}$ $\dfrac{2}{5} = \dfrac{2 \times 4}{5 \times 4} = \dfrac{8}{20}$

To simplify a fraction, divide both the numerator and the denominator by the same number, preferably the HCF (highest common factor).
Continue dividing until the only common factor in the numerator and the denominator is 1.

EXAMPLE 11

Simplify each fraction.

a $\dfrac{12}{15}$ **b** $\dfrac{16}{40}$

SOLUTION

a $\dfrac{12}{15} = \dfrac{12 \div 3}{15 \div 3} = \dfrac{4}{5}$

OR on a calculator, enter: 12 (a^b/c) 15 (=)

b $\dfrac{16}{40} = \dfrac{16 \div 8}{40 \div 8} = \dfrac{2}{5}$

OR enter: 16 (a^b/c) 40 (=)

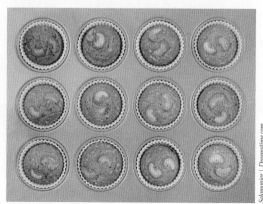

Solomonjee | Dreamstime.com

1 Change $\dfrac{8}{5}$ to a mixed numeral. Select the correct answer **A**, **B**, **C** or **D**.

   A $1\dfrac{3}{5}$ **B** $1\dfrac{1}{5}$ **C** $2\dfrac{3}{5}$ **D** $1\dfrac{3}{8}$

2 Change $3\dfrac{1}{4}$ to an improper fraction. Select **A**, **B**, **C** or **D**.

   A $\dfrac{12}{4}$ **B** $\dfrac{4}{13}$ **C** $\dfrac{13}{4}$ **D** $\dfrac{8}{4}$

3 Which of these fractions are improper fractions?

   $\dfrac{9}{4}$ $2\dfrac{1}{2}$ $\dfrac{3}{8}$ $\dfrac{11}{5}$ $1\dfrac{4}{5}$ $\dfrac{9}{7}$ $\dfrac{7}{5}$ $2\dfrac{3}{5}$

4 Change each improper fraction to a mixed numeral.

   a $\dfrac{9}{4}$ **b** $\dfrac{5}{2}$ **c** $\dfrac{9}{5}$ **d** $\dfrac{15}{8}$

   e $\dfrac{5}{3}$ **f** $\dfrac{13}{7}$ **g** $\dfrac{13}{6}$ **h** $\dfrac{15}{9}$

5 Change each mixed numeral to an improper fraction.

   a $1\dfrac{1}{3}$ **b** $3\dfrac{1}{4}$ **c** $4\dfrac{3}{5}$ **d** $1\dfrac{5}{8}$

   e $2\dfrac{2}{5}$ **f** $3\dfrac{5}{6}$ **g** $1\dfrac{5}{9}$ **h** $2\dfrac{4}{5}$

6 Jackie bought 7 cakes and cut them into 8 slices each. Her friends ate $4\dfrac{5}{8}$ cakes.

   a How many pieces did they eat?

   b How many pieces were left?

7 Copy and complete each pair of equivalent fractions.

   a $\dfrac{2}{3} = \dfrac{\quad}{12}$ **b** $\dfrac{4}{5} = \dfrac{20}{\quad}$ **c** $\dfrac{\quad}{3} = \dfrac{10}{15}$ **d** $\dfrac{5}{\quad} = \dfrac{35}{56}$

8 Is each equation true or false?

   a $\dfrac{6}{10} = \dfrac{3}{5}$ **b** $\dfrac{12}{18} = \dfrac{3}{2}$ **c** $\dfrac{15}{25} = \dfrac{3}{5}$ **d** $\dfrac{30}{55} = \dfrac{5}{11}$

9 Simplify each fraction and check your answer on a calculator.

   a $\dfrac{6}{20}$ **b** $\dfrac{15}{45}$ **c** $\dfrac{16}{24}$ **d** $\dfrac{12}{27}$

   e $\dfrac{10}{26}$ **f** $\dfrac{12}{20}$ **g** $\dfrac{70}{100}$ **h** $\dfrac{24}{60}$

   i $\dfrac{32}{56}$ **j** $\dfrac{25}{75}$

WORDBANK

ascending order From smallest to largest, to go **up** in size.

descending order From largest to smallest, to go **down** in size.

< 'is less than'

> 'is greater than'

lowest common multiple (LCM) The smallest multiple of two or more numbers, for example, the LCM of 4 and 6 is 12.

To **order or compare fractions**, we must first use equivalent fractions to write them with the **same denominator**, then compare their **numerators**.

EXAMPLE 12

Copy and complete each statement using the signs <, > or =.

a $\dfrac{3}{5}\underline{\quad}\dfrac{2}{3}$ **b** $\dfrac{5}{8}\underline{\quad}\dfrac{4}{7}$ **c** $\dfrac{3}{4}\underline{\quad}\dfrac{9}{12}$

SOLUTION

a Make both fractions have the same denominator.

For denominators 5 and 3, a common denominator is $5 \times 3 = 15$.

$\dfrac{3}{5} = \dfrac{9}{15}$ and $\dfrac{2}{3} = \dfrac{10}{15}$

So $\dfrac{3}{5} < \dfrac{2}{3}$ ⟵ because $9 < 10$

b For denominators 7 and 8, a common denominator is $7 \times 8 = 56$.

$\dfrac{5}{8} = \dfrac{35}{56}$ and $\dfrac{4}{7} = \dfrac{32}{56}$

So $\dfrac{5}{8} > \dfrac{4}{7}$ ⟵ because $35 > 32$

c For denominators 4 and 12, the lowest common multiple (LCM) is 12.

$\dfrac{3}{4} = \dfrac{9}{12}$ and $\dfrac{9}{12} = \dfrac{9}{12}$

So $\dfrac{3}{4} = \dfrac{9}{12}$.

EXAMPLE 13

Write these fractions in ascending order: $\dfrac{5}{8}, \dfrac{3}{4}, \dfrac{5}{12}, \dfrac{5}{6}$.

SOLUTION

Make all fractions have the same denominator, 24.

$\dfrac{5}{8} = \dfrac{15}{24}, \dfrac{3}{4} = \dfrac{18}{24}, \dfrac{5}{12} = \dfrac{10}{24}, \dfrac{5}{6} = \dfrac{20}{24}.$

In ascending order: $\dfrac{10}{24}, \dfrac{15}{24}, \dfrac{18}{24}, \dfrac{20}{24}$ ⟵ from smallest to largest

which is: $\dfrac{5}{12}, \dfrac{5}{8}, \dfrac{3}{4}, \dfrac{5}{6}$

1 Which fraction is the smallest? Select the correct answer **A, B, C** or **D**.

   A $\dfrac{2}{3}$ **B** $\dfrac{1}{2}$ **C** $\dfrac{3}{5}$ **D** $\dfrac{3}{4}$

2 Which fraction in Question **1** is the largest? Select **A, B, C** or **D**.

3 Is each statement true or false?

   a $\dfrac{1}{4} < \dfrac{2}{5}$ **b** $\dfrac{3}{5} > \dfrac{3}{8}$

   c $\dfrac{8}{12} = \dfrac{2}{3}$ **d** $\dfrac{2}{5} < \dfrac{3}{12}$

   e $\dfrac{4}{7} > \dfrac{3}{8}$ **f** $\dfrac{3}{8} = \dfrac{10}{24}$

   g $\dfrac{5}{8} > \dfrac{10}{12}$ **h** $\dfrac{5}{6} < \dfrac{7}{8}$

4 List these fractions in ascending order: $\dfrac{2}{3}, \dfrac{3}{24}, \dfrac{5}{6}, \dfrac{9}{12}$

5 List these fractions in descending order: $\dfrac{3}{4}, \dfrac{7}{10}, \dfrac{13}{20}, \dfrac{3}{5}$

6 Copy and complete each statement using the signs <, > or =.

   a $\dfrac{1}{2} \underline{\quad} \dfrac{2}{3}$ **b** $\dfrac{3}{4} \underline{\quad} \dfrac{5}{8}$

   c $\dfrac{6}{7} \underline{\quad} \dfrac{19}{21}$ **d** $\dfrac{5}{8} \underline{\quad} \dfrac{7}{12}$

   e $\dfrac{1}{7} \underline{\quad} \dfrac{1}{8}$ **f** $\dfrac{6}{5} \underline{\quad} \dfrac{9}{8}$

   g $\dfrac{2}{5} \underline{\quad} \dfrac{5}{2}$ **h** $\dfrac{3}{5} \underline{\quad} \dfrac{9}{15}$

7 Copy each number line, then place each fraction in the correct position.

   a $\dfrac{1}{2} \quad \dfrac{3}{4} \quad \dfrac{5}{8} \quad \dfrac{1}{4} \quad \dfrac{7}{8}$

   b $\dfrac{1}{2} \quad \dfrac{3}{4} \quad \dfrac{2}{3} \quad \dfrac{5}{6} \quad \dfrac{5}{12}$

8 Mia had a box of chocolates and gave Talia $\dfrac{3}{8}$ of the chocolates and Brendon $\dfrac{5}{12}$ of them. Who had the greater share?

WORDBANK

lowest common denominator (LCD) The lowest common multiple (LCM) of the denominators of two or more fractions; for example, the LCD of $\frac{2}{3}$ and $\frac{5}{6}$ is 6.

To add or subtract fractions:
- convert them (if needed) so that they have the same denominator, preferably the **lowest common denominator (LCD)**
- add or subtract the **numerators** and keep the denominators the same.

EXAMPLE 14

Evaluate each expression.

a $\frac{1}{4}+\frac{2}{3}$

b $\frac{4}{5}+\frac{7}{15}$

c $3+1\frac{3}{8}$

d $\frac{5}{6}-\frac{3}{8}$

e $\frac{17}{18}-\frac{1}{6}$

f $5-\frac{3}{7}$

SOLUTION

a $\frac{1}{4}+\frac{2}{3}=\frac{3}{12}+\frac{8}{12}$

 ✳ LCD of 4 and 3 is 12.

$=\frac{11}{12}$

OR on a calculator, enter:

$1\ \boxed{a^{b}/_{c}}\ 4\ \boxed{+}\ 2\ \boxed{a^{b}/_{c}}\ 3\ \boxed{=}$

b $\frac{4}{5}+\frac{7}{15}=\frac{12}{15}+\frac{7}{15}$

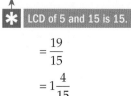 ✳ LCD of 5 and 15 is 15.

$=\frac{19}{15}$

$=1\frac{4}{15}$

c $3+1\frac{3}{8}=3+1+\frac{3}{8}$

$=4\frac{3}{8}$

d $\frac{5}{6}-\frac{3}{8}=\frac{20}{24}-\frac{9}{24}$

$=\frac{11}{24}$

e $\frac{17}{18}-\frac{1}{6}=\frac{17}{18}-\frac{3}{18}$

$=\frac{14}{18}$

$=\frac{7}{9}$

f $5-\frac{3}{7}=4+1-\frac{3}{7}$

$=4+\frac{4}{7}$

$=4\frac{4}{7}$

1 What is the lowest common denominator of $\dfrac{2}{3}$ and $\dfrac{1}{4}$? Select the correct answer **A, B, C** or **D**.

    A 3   **B** 4   **C** 24   **D** 12

2 What is the LCD of $\dfrac{1}{6}$ and $\dfrac{5}{9}$? Select **A, B, C** or **D**.

    A 12   **B** 9   **C** 18   **D** 36

3 Copy and complete each line of working.

    a $\dfrac{1}{3} = \dfrac{}{15}$ **b** $\dfrac{3}{5} = \dfrac{}{15}$ **c** $\dfrac{1}{3} + \dfrac{3}{5} = \dfrac{}{15} + \dfrac{}{15}$

$$= \dfrac{}{15}$$

    d $\dfrac{2}{3} = \dfrac{}{12}$ **e** $\dfrac{1}{4} = \dfrac{}{12}$ **f** $\dfrac{2}{3} - \dfrac{1}{4} = \dfrac{}{12} - \dfrac{}{12}$

$$= \dfrac{}{12}$$

4 Evaluate each sum, then use your calculator to check your answers.

    a $\dfrac{1}{2} + \dfrac{1}{4}$ **b** $\dfrac{1}{6} + \dfrac{1}{3}$ **c** $\dfrac{1}{3} + \dfrac{3}{5}$ **d** $\dfrac{3}{8} + \dfrac{1}{6}$

    e $\dfrac{3}{5} + \dfrac{1}{8}$ **f** $\dfrac{2}{9} + \dfrac{3}{4}$ **g** $2 + \dfrac{4}{5}$ **h** $\dfrac{7}{8} + \dfrac{13}{4}$

    i $2\dfrac{2}{5} + 4$ **j** $2\dfrac{1}{4} + \dfrac{2}{3}$

5 Evaluate each difference, then use your calculator to check your answers.

    a $\dfrac{5}{8} - \dfrac{1}{2}$ **b** $\dfrac{4}{5} - \dfrac{2}{3}$ **c** $\dfrac{7}{8} - \dfrac{1}{4}$ **d** $\dfrac{8}{9} - \dfrac{2}{3}$

    e $\dfrac{7}{12} - \dfrac{1}{4}$ **f** $\dfrac{7}{9} - \dfrac{2}{3}$ **g** $5 - \dfrac{2}{3}$ **h** $\dfrac{17}{18} - \dfrac{5}{6}$

    i $2\dfrac{3}{4} - 1\dfrac{1}{2}$ **j** $4\dfrac{3}{4} - \dfrac{2}{3}$

6 Use your calculator to evaluate each expression.

    a $2\dfrac{1}{2} + 3\dfrac{1}{5}$ **b** $3\dfrac{3}{5} - 2\dfrac{1}{3}$ **c** $5\dfrac{3}{8} + 2\dfrac{2}{5}$

    d $6\dfrac{5}{8} - 1\dfrac{3}{4}$ **e** $1\dfrac{5}{12} + 3\dfrac{1}{2} - 2\dfrac{1}{3}$

7 The ingredients of a sponge cake are $\dfrac{1}{6}$ sugar, $\dfrac{3}{8}$ flour, $\dfrac{1}{4}$ eggs and the rest is made up of milk and butter.

    a What fraction of the ingredients are sugar, flour and eggs combined?

    b Find the fraction left for the milk and butter.

    c If I cut the cake into 16 pieces and Zina eats $\dfrac{1}{8}$ of it, how many pieces does she eat?

WORDBANK

reciprocal The reciprocal of a fraction is the fraction turned upside down. For example, the reciprocal of $\frac{2}{3}$ is $\frac{3}{2}$.

To multiply fractions:
- simplify numerators with denominators (if possible) by dividing by a common factor
- then **multiply numerators** and **multiply denominators**

EXAMPLE 15

Evaluate each product.

a $\dfrac{2}{3} \times \dfrac{4}{7}$ **b** $\dfrac{3}{4} \times \dfrac{12}{17}$ **c** $\dfrac{5}{8} \times 10$

SOLUTION

a $\dfrac{2}{3} \times \dfrac{4}{7} = \dfrac{2 \times 4}{3 \times 7}$

$= \dfrac{8}{21}$

OR on a calculator, enter:

b $\dfrac{3}{4} \times \dfrac{12}{17} = \dfrac{3 \times \cancel{12}^{3}}{{}_{1}\cancel{4} \times 17}$

$= \dfrac{9}{17}$

c $\dfrac{5}{8} \times 10 = \dfrac{5 \times \cancel{10}^{5}}{\cancel{8}^{4}}$

$= \dfrac{25}{4}$

$= 6\dfrac{1}{4}$

To divide by a fraction $\dfrac{a}{b}$, multiply by its reciprocal $\dfrac{b}{a}$.

EXAMPLE 16

Evaluate each quotient.

a $\dfrac{3}{5} \div \dfrac{7}{10}$ **b** $\dfrac{5}{8} \div \dfrac{15}{4}$ **c** $\dfrac{2}{9} \div 6$

SOLUTION

a $\dfrac{3}{5} \div \dfrac{7}{10} = \dfrac{3}{5} \times \dfrac{10}{7}$

$= \dfrac{3 \times \cancel{10}^{2}}{{}_{1}\cancel{5} \times 7}$

$= \dfrac{6}{7}$

OR on a calculator, enter:

b $\dfrac{5}{8} \div \dfrac{15}{4} = \dfrac{5}{8} \times \dfrac{4}{15}$

$= \dfrac{{}^{1}\cancel{5} \times \cancel{4}^{1}}{{}^{2}\cancel{8} \times \cancel{15}^{3}}$

$= \dfrac{1}{6}$

c $\dfrac{2}{9} \div 6 = \dfrac{2}{9} \times \dfrac{1}{6}$

※ The reciprocal of 6 is $\dfrac{1}{6}$

$= \dfrac{{}^{1}\cancel{2} \times 1}{9 \times \cancel{6}^{3}}$

$= \dfrac{1}{27}$

1   Evaluate the product of $\frac{1}{5}$ and $\frac{3}{4}$. Select the correct answer **A, B, C** or **D**.

   A $\frac{3}{9}$ **B** $\frac{3}{12}$ **C** $\frac{1}{20}$ **D** $\frac{3}{20}$

2   Evaluate $\frac{2}{5} \div \frac{4}{3}$. Select **A, B, C** or **D**.

   A $\frac{3}{10}$ **B** $\frac{8}{15}$ **C** 1   **D** $\frac{20}{6}$

3   Copy and complete this equation.

   $$\frac{2}{7} \times \frac{3}{5} = \frac{\underline{\quad} \times 3}{7 \times \underline{\quad}} = \frac{\overline{\quad\quad}}{\overline{\quad\quad}}$$

4   Evaluate each product, then check your answers on a calculator.

   a $\frac{1}{5} \times \frac{1}{3}$ **b** $\frac{2}{3} \times \frac{3}{8}$ **c** $\frac{3}{8} \times \frac{1}{4}$ **d** $\frac{5}{12} \times \frac{5}{8}$

   e $\frac{4}{5} \times \frac{5}{6}$ **f** $\frac{5}{8} \times \frac{4}{9}$ **g** $\frac{4}{9} \times 27$ **h** $\frac{7}{9} \times \frac{3}{21}$

   i $\frac{2}{5} \times 16$ **j** $\frac{3}{20} \times \frac{5}{18}$

5   Copy and complete this equation.

   $$\frac{2}{3} \div \frac{3}{8} = \frac{\overline{\quad\quad}}{3} \times \frac{\overline{\quad\quad}}{3} = \frac{\overline{\quad\quad}}{\underline{\quad}} = 1\frac{\overline{\quad\quad}}{\underline{\quad}}$$

6   Evaluate each quotient, then check your answers on a calculator.

   a $\frac{3}{4} \div \frac{1}{2}$ **b** $\frac{2}{5} \div \frac{4}{5}$ **c** $\frac{3}{8} \div \frac{1}{4}$ **d** $\frac{7}{12} \div \frac{9}{8}$

   e $\frac{4}{3} \div \frac{16}{3}$ **f** $\frac{1}{6} \div \frac{4}{9}$ **g** $\frac{4}{9} \div 12$ **h** $\frac{5}{8} \div \frac{15}{24}$

   i $\frac{3}{5} \div 15$ **j** $\frac{6}{15} \div \frac{12}{5}$

7   Use your calculator to evaluate each expression.

   a $2 \times 1\frac{2}{7}$ **b** $3\frac{1}{8} \times 2\frac{1}{2}$ **c** $4\frac{3}{8} \div 3$

   d $5\frac{3}{8} \div 1\frac{3}{4}$ **e** $3\frac{1}{4} \times 2 \times 1\frac{3}{8}$

8   Joel was running a business and worked out that the wages were equal to $\frac{1}{3}$ of the takings (income) for a week and the profit was $\frac{1}{4}$ of the wages.

   a   What fraction of the takings was the profit?

   b   If the takings for 1 week were \$28 500, how much were the wages?

   c   What was the profit for the week?

WORD PUZZLE

List the letters in order for the clues below to spell out a phrase relating to this topic.

The first $\dfrac{3}{7}$ of FRANTIC

The first $\dfrac{1}{6}$ of CAREER

The last $\dfrac{4}{10}$ of CORRECTION

The first $\dfrac{2}{7}$ of DISTANT

The first $\dfrac{1}{3}$ of STRENUOUS

The second $\dfrac{1}{4}$ of REACTION

The first $\dfrac{2}{7}$ of TITANIC

The last $\dfrac{1}{4}$ of CREATION

Part A General topics

Calculators are not allowed.

1 Evaluate 50×9.

2 Copy and complete: 45 km = _____ m

3 List the factors of 12.

4 Simplify $-2x + 3y - y + 5x$

5 Find the range of the scores 14, 5, 7, 9, 11, 7, 10.

6 Given that $18 \times 6 = 108$, evaluate 1.8×6.

7 Find 5% of $360.

8 Find the value of d if the perimeter of this rectangle is 84 m.

9 Simplify $4 \times x \times 7 \times y$

10 Copy and complete: $\dfrac{3}{8} = \dfrac{}{32}$.

Part B Working with numbers

Calculators are allowed.

1–01 Mental addition and subtraction

11 Which equation is correct? Select the correct answer **A**, **B**, **C** or **D**.

    A $52 + 11 = 52 + 10 - 1$ **B** $52 + 9 = 52 + 10 - 1$

    C $52 - 11 = 52 + 10 - 1$ **D** $52 - 9 = 52 + 10 - 1$

12 Evaluate each difference.

    a $65 - 9$ **b** $182 - 31$ **c** $2189 - 16$

1–02 Mental multiplication and division

13 Which expression gives the same answer as 50×12? Select **A**, **B**, **C** or **D**.

    A $50 \times 6 \times 6$ **B** $50 \times 10 + 50 \times 2$

    C $50 \times 10 \times 2$ **D** $100 \times 2 \times 6$

14 Evaluate each quotient.

    a $240 \div 5$ **b** $380 \div 20$ **c** $7264 \div 8$

1–03 Integers

15 Evaluate each expression.

    a $-6 + 3$ **b** $-7 + (-4)$ **c** $-12 - (-7)$

    d $-3 \times (-9)$ **e** $72 \div (-8)$ **f** $8 \times (-3) \times (-2)^2$

1–04 Order of operations

16 Evaluate each expression.

    a $32 + (-3) \times 5$ **b** $24 - (-8) \div 2$ **c** $6 \times (-3) + (-20) \div 4$

1-05 Fractions

17 Change each improper fraction to a mixed numeral.

    a $\dfrac{7}{4}$ **b** $\dfrac{13}{6}$ **c** $\dfrac{34}{9}$

1-06 Ordering fractions

18 Is each statement true or false?

    a $\dfrac{1}{3} < \dfrac{3}{5}$ **b** $\dfrac{3}{5} > \dfrac{7}{8}$ **c** $\dfrac{3}{4} = \dfrac{14}{20}$

1-07 Adding and subtracting fractions

19 Evaluate each expression.

    a $\dfrac{3}{4} + \dfrac{1}{3}$ **b** $\dfrac{5}{12} - \dfrac{1}{8}$ **c** $1\dfrac{1}{2} + \dfrac{1}{4}$

1-08 Multiplying and dividing fractions

20 Evaluate each expression.

    a $\dfrac{4}{5} \times \dfrac{3}{8}$ **b** $\dfrac{5}{9} \div \dfrac{3}{18}$ **c** $2\dfrac{1}{2} \times \dfrac{4}{15}$

21 Rani made a cake and cut it in half. She then cut each piece into sixths.

    a How many pieces did she cut the cake into?

    b Nicky ate $2\dfrac{1}{2}$ pieces. What fraction of the cake did she eat?

PERCENTAGES

2

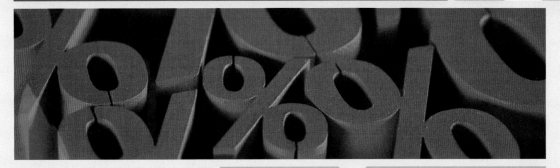

IN THIS CHAPTER YOU WILL:

- convert between percentages, fractions and decimals
- compare percentages, fractions and decimals
- find a percentage of a number or metric quantity
- express quantities as fractions and percentages of a whole
- calculate percentage increases and decreases
- use the unitary method to find a whole amount given a percentage of it
- solve problems involving profit and loss, cost price and selling price
- solve problems involving discounts and GST

Shutterstock.com/Sebastian Duda

| Percentages, fractions and decimals

A **percentage** is a fraction with a denominator of 100; for example, 30% means 30 out of 100 or $\dfrac{30}{100}$.

EXAMPLE 1

Convert each percentage to a fraction and a decimal.

a 35%   **b** 18.6%   **c** $33\dfrac{1}{3}\%$

SOLUTION

a $35\% = \dfrac{35}{100}$

$= \dfrac{7}{20}$

b $18.6\% = \dfrac{18.6}{100}$

$= \dfrac{18.6 \times 10}{100 \times 10}$

$= \dfrac{186}{1000}$

$= \dfrac{93}{500}$

c $33\dfrac{1}{3}\% = \dfrac{33\frac{1}{3}}{100}$

$= \dfrac{33\frac{1}{3} \times 3}{100 \times 3}$

$= \dfrac{100}{300}$

$= \dfrac{1}{3}$

$35\% = 35 \div 100$
$= 0.35$

$18.6\% = 18.6 \div 100$
$= 0.186$

$33\dfrac{1}{3}\% = 33\dfrac{1}{3} \div 100$
$= 0.3333\ldots$
$= 0.\dot{3}$

EXAMPLE 2

Convert each number to a percentage.

a $\dfrac{3}{5}$ **b** 0.8   **c** 1.75

SOLUTION

a $\dfrac{3}{5} = \dfrac{3}{5} \times 100\%$
$= 60\%$

b $0.8 = 0.8 \times 100\%$
$= 80\%$

c $1.75 = 1.75 \times 100\%$
$= 175\%$

To compare percentages, fractions and decimals:
▓ convert them all to decimals
▓ make them all have the same number of decimal places by adding zeros where necessary

---

EXAMPLE 3

Write $\dfrac{4}{5}$, 75% and 0.725 in ascending order.

SOLUTION

$\dfrac{4}{5} = 4 \div 5 = 0.8$ $75\% = 0.75$ 0.725

Write each decimal with 3 decimal places.

$\dfrac{4}{5} = 0.800$ $75\% = 0.750$ $0.725 = 0.725$

From smallest to largest: 0.725, 0.750, 0.800

So 0.725, 75%, $\dfrac{4}{5}$ are in ascending order.

---

1 Write 30% as a fraction in simplest form. Select the correct answer **A**, **B**, **C** or **D**.

   A $\dfrac{30}{1000}$ **B** $\dfrac{30}{100}$ **C** $\dfrac{3}{10}$ **D** $\dfrac{1}{3}$

2 Write 4.5% as a decimal. Select **A**, **B**, **C** or **D**.

   A 0.45   **B** 0.045   **C** 4.50   **D** 0.0045

3 Convert each percentage to a simplified fraction.

   a 2%   **b** 5%   **c** 13%   **d** 20%
   e 23%   **f** 30%   **g** 38%   **h** 45%
   i 62%   **j** 70%   **k** 84%   **l** 95%
   m 105%   **n** 120%   **o** 150%

4 Convert each percentage in Question **3** to a decimal.

5 Copy and complete each conversion from a fraction to a percentage.

   a $\dfrac{3}{4} = \dfrac{3}{4} \times 100\% =$ ___ %

   b $\dfrac{3}{8} = \dfrac{3}{8} \times$ ___ % = ____ %

6 Convert each fraction to a percentage.

   a $\dfrac{1}{4}$ **b** $\dfrac{2}{5}$ **c** $\dfrac{5}{8}$ **d** $\dfrac{3}{10}$

   e $\dfrac{5}{12}$ **f** $\dfrac{7}{10}$ **g** $\dfrac{5}{9}$ **h** $\dfrac{2}{3}$

   i $\dfrac{7}{8}$ **j** $\dfrac{13}{20}$

7 Convert each decimal to a percentage.

   a 0.6   **b** 0.02   **c** 0.4   **d** 0.18
   e 0.32   **f** 0.85   **g** 0.125 **h** 0.93
   i 1.15   **j** 2.5

8 Write each set of numbers in ascending order by converting them to decimals first.

   a $\dfrac{3}{8}, 0.34, 35\%$ **b** $\dfrac{7}{10}, 74\%, 0.705$ **c** $25\%, \dfrac{5}{18}, 0.255$

9 Write each set of numbers in descending order by converting them to decimals first.

   a $\dfrac{3}{4}, 0.745, 76\%$ **b** $\dfrac{3}{10}, 34\%, 0.315$ **c** $85\%, \dfrac{5}{6}, 0.845$

10 Copy and complete this table.

Fraction		$\dfrac{1}{10}$		$\dfrac{1}{4}$			$\dfrac{7}{10}$		$\dfrac{4}{5}$	
Decimal	0.05				0.4					1
Percentage			20%			60%		75%		

This table lists some commonly-used percentages for mental calculation.

Percentage	Fraction	Decimal
10%	$\dfrac{1}{10}$	0.1
12.5%	$\dfrac{1}{8}$	0.125
25%	$\dfrac{1}{4}$	0.25
$33\dfrac{1}{3}\%$	$\dfrac{1}{3}$	$0.\dot{3}$
50%	$\dfrac{1}{2}$	0.5
$66\dfrac{2}{3}\%$	$\dfrac{2}{3}$	$0.\dot{6}$
75%	$\dfrac{3}{4}$	0.75
80%	$\dfrac{4}{5}$	0.8
100%	1	1.0

EXAMPLE 4

Use mental calculation to find each quantity.

a 75% of $56

b $33\dfrac{1}{3}\%$ of 18 km

c 10% of 50 days

SOLUTION

a $75\% \text{ of } \$56 = \dfrac{3}{4} \times \$56$

$= \$42$

b $33\dfrac{1}{3}\% \text{ of } 18 \text{ km} = \dfrac{1}{3} \times 18$

$= 6 \text{ km}$

c $10\% \text{ of } 50 \text{ days} = \dfrac{1}{10} \times 50$

$= 5 \text{ days}$

EXAMPLE 5

Find each quantity.

a 28% of $480

b 55% of 25 L

c 2.5% of 20 years

SOLUTION

a $28\% \text{ of } \$480 = 0.28 \times \$480$

$= \$134.40$

⟵ or enter 28 `%` `×` 480 `=` on a calculator

b $55\% \text{ of } 25 \text{ L} = 0.55 \times 25 \text{ L}$

$= 13.75 \text{ L}$

⟵ or enter 55 `%` `×` 25 `=` on a calculator

c $2.5\% \text{ of } 20 \text{ years} = 0.025 \times 20 \times 12 \text{ months}$

$= 6 \text{ months}$

⟵ converting 20 years to months

1   Write 40% as a simplified fraction. Select the correct answer **A, B, C** or **D**.

   A $\dfrac{40}{100}$ **B** $\dfrac{40}{1000}$ **C** $\dfrac{4}{10}$ **D** $\dfrac{2}{5}$

2   Write 55% as a simplified fraction. Select **A, B, C** or **D**.

   A $\dfrac{55}{100}$ **B** $\dfrac{11}{20}$ **C** $\dfrac{5.5}{10}$ **D** $\dfrac{55}{1000}$

3   Write each percentage as a simplified fraction.

   a 25%   **b** 12.5%   **c** 80%   **d** 50%

   e $33\dfrac{1}{3}\%$ **f** 100%   **g** 10%   **h** $66\dfrac{2}{3}\%$

   i 75%   **j** 40%

4   Copy and complete the following.

   a 25% of $260 = $\dfrac{1}{-} \times 260$ **b** 80% of 200 cm = $\dfrac{-}{5} \times 200$

   $\qquad\qquad\qquad = \$ \underline{\quad}$ $\qquad\qquad\qquad = \underline{\quad}$ cm

5   Use mental calculation to find each quantity.

   a 10% of $800   **b** 25% of 8 years   **c** 50% of 64 km

   d 80% of 450 cm   **e** 75% of $2400   **f** 150% of 24 years

   g $33\dfrac{1}{3}\%$ of 1200 L   **h** $12\dfrac{1}{2}\%$ of 16 days   **i** $66\dfrac{2}{3}\%$ of 96 m

   j $12\dfrac{1}{2}\%$ of $840   **k** 75% of $44   **l** 75% of 84 km

   m 50% of $120   **n** 50% of $1.20   **o** 20% of $150

   p 20% of $1.50   **q** $66\dfrac{2}{3}\%$ of $42   **r** $66\dfrac{2}{3}\%$ of $4.20

6   Find each quantity.

   a 5% of $350   **b** 24% of 8 m   **c** 60% of 85 m

   d 12.5% of 56 years   **e** 62% of 180 cm   **f** 85% of 40 years

   g 15% of 12 L   **h** 45% of 1 day   **i** 22% of 10 km

   j 49% of $1260   **k** 3.5% of 25 days   **l** 160% of $18 000

7   There were 65 000 people at the concert and 35% of the audience were male.

   a How many in the audience were male?

   b How many were female?

8   If a swimming pool holds 72 litres of water and is only 12.5% full, how many
    litres of water would the pool hold when completely full?

9   Rhianna spent 88% of her holiday savings while overseas. If she had saved $8400,
    how much money did she have left?

To express an amount as a fraction of a whole amount, write the fraction as $\dfrac{\text{amount}}{\text{whole amount}}$ and simplify if possible.

EXAMPLE 6

Express each amount as a fraction.

a 32 marks out of 40

b 12 minutes out of $1\dfrac{1}{2}$ hours

SOLUTION

a 32 marks out of $40 = \dfrac{32}{40}$

$= \dfrac{4}{5}$

b 12 minutes out of 1.5 hours $= \dfrac{12}{90}$ ⟵ $1\dfrac{1}{2}$ hours = 90 min

$= \dfrac{2}{15}$

To express an amount as a percentage of a whole amount, calculate $\dfrac{\text{amount}}{\text{whole amount}} \times 100\%$.

EXAMPLE 7

Express each amount as a percentage.

a $150 out of $250

b 16 cm out of 4 m

SOLUTION

a $150 out of $250 $= \dfrac{150}{250} \times 100\%$

$= 60\%$

b 16 cm out of 4 m $= \dfrac{16}{400} \times 100\%$ ⟵ 4 m = 400 cm

$= 4\%$

 When expressing amounts as fractions or percentages, all units must be the same.

Paul Thompson Images / Alamy

1   Express 42 out of 60 as a simplified fraction. Select the correct answer **A**, **B**, **C** or **D**.

   A $\dfrac{42}{60}$ **B** $\dfrac{6}{10}$ **C** $\dfrac{14}{20}$ **D** $\dfrac{7}{10}$

2   Express 36 out of 90 as a percentage. Select **A**, **B**, **C** or **D**.

   A 36%   **B** 40%   **C** 30%   **D** 4%

3   Copy and complete each statement to write the answer as a fraction.

   a 15 out of 40 $= \dfrac{15}{} = -$ **b** 80 out of 220 $= \dfrac{}{220} = -$

4   Express each amount as a fraction.

   a 30 out of 50   **b** 18 mins out of 1 hour

   c 25 cm out of 1 m   **d** $6 out of $10

   e 36 mL out of 1 L   **f** 25c out of $4

   g 3 months of 1 year   **h** 16 m out of 2 km

   i 8 hours out of 2 days

5   Express each amount in Question **4** as a percentage.

6   Copy and complete each statement to write the answer as a percentage.

   a 18 girls out of 40 students $= \dfrac{18}{} \times 100\%$ **b** 50 out of 80 $= \dfrac{}{80} \times 100\%$

   $ = \underline{}\%$ $ = \underline{}\%$

7   Express each amount as a percentage.

   a 24 out of 60   **b** 15 out of 40

   c 36 out 60   **d** 50c out of $5

   e 12 cm out of 2 m   **f** 8 hours out of 1 day

   g 50 mL out of 2 L   **h** 75c out of $2

   i 48 mm out of 60 cm

8   When Justin went to bed, the temperature was 20°C. He woke up at 3 a.m. and the temperature had dropped by 4°.

   a What percentage decrease in temperature was this?

   b When Justin woke in the morning, the temperature had risen 6° since 3 a.m. What was the new temperature?

   c What percentage increase was this compared to the original temperature?

9   Convert each mark below to a percentage and then order the students' performances from best to worst.

   Blake: 50 out of 75   Grace: 60 out of 80   Kurt: 35 out of 50

   Tahlia: 75 out of 90   Jayden: 55 out of 60   Isabella: 98 out of 100

WORDBANK

increase　To make larger.

decrease　To make smaller.

To increase an amount by a percentage, find the percentage of the amount and add it to the amount.
To decrease an amount by a percentage, find the percentage of the amount and subtract it from the amount.

EXAMPLE 8

a　Increase $250 by 15%.

b　Decrease 60 km by 8%.

SOLUTION

a　Increase = 15% of $250
　　　　　 = $37.50

　　Increased amount = $250 + $37.50
　　　　　　　　　　 = $287.50

b　Decrease = 8% of 60 km
　　　　　 = 4.8 km

　　Decreased amount = 60 − 4.8 km
　　　　　　　　　　 = 55.2 km

EXAMPLE 9

At a sale, a TV is reduced by 25% from its recommended retail price of $1148.
What is the sale price of the TV?

SOLUTION

Decrease = 25% of $1148
　　　　 = $287

Sale price = $1148 − $287
　　　　　 = $861

British Retail Photography / Alamy

1   Increase $600 by 30%. Select the correct answer **A, B, C** or **D**.

    A $180   **B** $120   **C** $420   **D** $780

2   Decrease 180 m by 25%. Select **A, B, C** or **D**.

    A 45 m   **B** 135 m   **C** 225 m   **D** 145 m

3   Copy and complete these sentences.

    To **increase** an amount by a percentage, find the percentage first and then _____ it to the existing amount. To **decrease** an amount by a percentage, find the percentage first and then _____ it from the existing amount.

4   Increase:

    a $200 by 8%   **b** 1200 m by 25%

    c 4 days by 12.5%   **d** 5000 km by 5%

    e $6400 by 75%   **f** 120 L by $33\frac{1}{3}$%

5   Decrease:

    a $2500 by 20%   **b** 180 mins by 60%

    c 7200 km by 37.5%   **d** 2400 kg by 25%

    e $620 by 18%   **f** 880 ha by 12.5%

6   Find the sale price of each item if the discount is 15% for all items in Jolly John's store.

    a tablet device $580

    b home theatre $750

    c computer $1640

7   Find the new sale price of each item if an increase of 5% is applied.

    a shirt $85   **b** pants $79.95   **c** shoes $120

8   Jordan earns $850 per week. His manager has given him a pay rise of 2.5%.

    a What is his Jordan's new weekly wage?

    b What are his new annual earnings?

9   Olivia opened a shoe shop and added 10% to the cost price of all her shoes. After 2 weeks she had not been selling much stock so she reduced the sale price of her shoes by 10%. A red pair of women's shoes has a cost price of $110.

    a Calculate the original sale price of these shoes.

    b Calculate the discounted price of these shoes. Is this the same as the cost price?

10 Increase $15 000 by 20% and then decrease the new amount by 20%. Why isn't the final answer $15 000? Discuss this in class.

WORDBANK

unit Unit means one or each. $6 per unit means $6 for one.

unitary method A method to find a unit amount and then use this amount to find the total amount.

To use the unitary method:
- use the given amount to find 1%
- multiply 1% by 100 to find the total amount (100%).

EXAMPLE 10

Ryan donates 8% of his Christmas bonus to his local hospital each year. How much was his Christmas bonus if he donated $280 to the hospital?

SOLUTION

8% of his Christmas bonus = $280

1% of his Christmas bonus = $280 ÷ 8   ⟵ Calculating 1% first.
$$= \$35$$

100% of his Christmas bonus = $35 × 100 ⟵ Multiplying by 100.
$$= \$3500$$

So Ryan's Christmas bonus was $3500.   ⟵ Check: 8% × $3500 = $280

EXAMPLE 11

Georgia pays 22.5% of her wage in tax. If she pays $268.50 per week in tax, how much does she earn per week before her tax is paid?

SOLUTION

22.5% of Georgia's wage = $268.50

1% of Georgia's wage = $268.50 ÷ 22.5   ⟵ Calculate 1% first
$$= \$11.9333...$$

100% of Georgia's wage = $11.9333... × 100 ⟵ Multiplying by 100
$$= \$1193.33...$$
$$\approx \$1193.33$$ ⟵ Rounding to the nearest cent.

So Georgia's weekly wage is $1193.33.   ⟵ Check: 22.5% × $1193.33 ≈ $268.50

✳ Answers to money problems should be rounded to the nearest cent.

1 If 25% of an amount is $65, what is the amount? Select the correct answer A, B, C or D.

    A $260   B $650   C $250   D $265

2 If 3% of an amount is $180, what is the amount? Select A, B, C or D.

    A $600   B $6000   C $60   D $1800

3 Is each statement true or false?

    a If 10% = $360, then 1% = $36

    b If 55% = $1100, then 1% = $2

    c If 60% = $240, then 1% = $40

4 Copy and complete each solution.

    a 25% of an amount = $450   b 40% of an amount = $560

        1% of the amount = $450 ÷ ___   1% of the amount = $560 ÷ ___

                        = $____   = $____

       100% of the amount = $____ × 100   100% of the amount = $____ × 100

                        = $_____   = $_____

5 Find each total amount if:

    a 5% of it is $90   b 10% of it is $60

    c 25% of it is $300   d 12% of it is 720 m

    e 64% of it is 1280 L   f 80% of it is 168 m

    g 25% of it is 4 hrs   h 7% of it is $2100

    i 44% of it is 880 kg

6 Ganesh scored 75 runs in a game of cricket. This was 15% of his team's runs. How many runs did Ganesh's team score?

7 Brooke paid $120 to have her gold chain valued. This was 2.5% of the chain's value. How much is her gold chain worth?

8 The town of Barangee has 120 people who cannot speak English. This is 6% of the town's population. How many people live in Barangee?

9 Adam pays 26.5% of his wage in PAYG tax each week.

    a If Adam's tax is $296.80, how much is his wage?

    b How can you check if this amount is correct?

10 Melissa spent 45% of her savings at a factory outlet store by midday.

    a If she had spent $67.50 by midday, how much were her savings?

    b How much money did Melissa have left?

WORDBANK

cost price The price at which an item was bought by a retailer

selling price The price at which an item was sold by a retailer

profit To sell an item at a higher price; profit = selling price – cost price

loss To sell an item at a lower price; loss = cost price – selling price

GST Goods and services tax charged by the government on most goods and services bought.

EXAMPLE 12

A watch was bought for $24 and sold for $36.

a Find the profit.

b Calculate the profit as a percentage of the selling price.

SOLUTION

a Profit = $36 – $24 = $12

b Profit as a percentage of the selling price = $\dfrac{12}{36} \times 100\%$ ⟵ $\dfrac{\text{profit}}{\text{selling price}} \times 100\%$

$$= 33\frac{1}{3}\%$$

EXAMPLE 13

A pair of headphones were bought for $64 and sold for $50.

a Find the loss.

b Calculate the loss as a percentage of the cost price.

SOLUTION

a Loss = $64 – $50 = $14

b Loss as a percentage of the cost price = $\dfrac{14}{64} \times 100\%$ ⟵ $\dfrac{\text{loss}}{\text{cost price}} \times 100\%$

$$= 21\frac{7}{8}\%$$

EXAMPLE 14

Lachlan bought a gym set worth $340 at a sale. 10% GST was added to the price, but he was then given a 25% discount. How much did Lachlan pay for the gym set?

SOLUTION

Price of gym set including GST = $340 + 10% of $340 ⟵ Add the GST
$$= \$374$$

Discount = 25% of $374 = $93.50

Price paid = $374 – $93.50 = $280.50

1   A cricket ball was bought for $38 and sold for $45. What was the profit or loss? Select the correct answer **A, B, C** or **D**.

   A profit $7   **B** loss $7   **C** profit $8   **D** loss $8

2   A tent was bought for $212 and sold for $205. What was the profit or loss? Select **A, B, C** or **D**.

   A profit $7   **B** loss $7   **C** profit $8   **D** loss $8

3   Copy and complete this table.

Cost price	Selling price	Profit or loss
$18.00	$26.00	$8.00 profit
$25.00	$38.00	
$50.00	$36.00	
$350.00		$60.00 loss
	$760.00	$35.00 profit
$1290.60		$108.50 loss

4   Jamal bought refrigerators for $800 and sold them for $1200.

   a   What profit did he make on each refrigerator?

   b   Find the profit as a percentage of the cost price.

   c   Find the profit as a percentage of the selling price.

5   Lauren bought apples from the market at 25c each and sold them for 40c each. Find her profit as a percentage of:

   a   the cost price   **b**   the selling price

6   Cameron bought old motor bikes for $750, rebuilt them and then sold them for $1250.

   a   What profit did he make on each bike?

   b   Find the profit as a percentage of the cost price.

   c   Find the profit as a percentage of the selling price.

7   The following items are for sale at Wurst and Moore, but 10% GST has to be added to find the selling price. Calculate the GST and the selling price for each item.

   a   Dress $120   **b**   Shirt $55   **c**   Trousers $75   **d**   Shoes $140

   e   Belt $32   **f**   Scarf $26   **g**   Necklace $48   **h**   Tie $36

8   During a clearance sale, each item in Question **7** has its selling price discounted by 30%. Find the discount price of each item in Question **7**.

9   Reece bought a new phone and traded in his old phone for a $50 discount off the price. The listed price of the new phone was $420 plus 10% GST. How much did Reece pay for his new phone?

FIND-A-WORD PUZZLE

Make a copy of this page, then find all the words listed below in this grid of letters.

C	R	E	G	A	T	N	E	C	R	E	P
O	E	E	C	I	R	P	C	O	S	T	N
M	T	S	E	V	W	C	O	N	L	A	O
M	A	Y	S	I	T	B	U	V	P	M	I
I	I	T	A	L	D	I	T	E	C	O	T
S	N	I	E	S	E	P	W	R	T	U	A
I	E	T	R	T	C	A	N	T	O	N	C
O	R	N	C	P	R	O	F	I	T	T	I
N	R	A	N	Y	E	B	E	S	S	O	L
E	Q	U	I	V	A	L	E	N	T	O	P
W	A	Q	U	E	S	T	R	A	D	E	P
S	L	A	P	S	E	L	L	I	N	G	A

AMOUNT	CONVERT	COST	DECREASE
EQUIVALENT	INCREASE	LOSS	PERCENTAGE
PROFIT	PRICE	QUANTITY	SELLING

ISBN 9780170351058

PRACTICE TEST 2

Part A General topics

Calculators are not allowed.

1 Evaluate: $22 - 3^2 \times 2$

2 List all of the factors of 6.

3 Given that $15 \times 6 = 90$, evaluate 1.5×6.

4 Find 75% of $400.

5 Simplify $\dfrac{18}{45}$.

6 Write these decimals in ascending order:
8.8, 8.89, 8.819.

7 What is the probability of a person chosen at random having a birthday in a month beginning with A?

8 Find the value of d if the perimeter of this rectangle is 68 m.

9 Simplify $-2 \times x \times 8 \times y^2$

10 Find the range of these scores:
12, 10, 9, 15, 17, 13.

Part B Pecentages

Calculators are allowed.

2–01 Percentages, fractions and decimals

11 Convert 30% to a decimal. Select the correct answer **A, B, C** or **D**.

  A 0.03   **B** 0.3   **C** 3.00   **D** 0.003

12 Convert $\dfrac{1}{5}$ to a percentage. Select **A, B, C** or **D**.

  A 5%   **B** 50%   **C** 25%   **D** 20%

2–02 Percentage of a quantity

13 Find 25% of $1800. Select **A, B, C** or **D**.

  A $450   **B** $360   **C** $4500   **D** $3600

14 Find each quantity.

  a 20% of 55 m   **b** $12\dfrac{1}{2}$% of 16 hours   **c** 80% of 65 L

2–03 Expressing quantities as fractions and percentages

15 Write each amount as a fraction.

  a 24 marks out of 30

  b 20 minutes out of 1 hour

  c 30 cm out of 4 m

16 Write each amount in Question **15** as a percentage.

2-04 Percentage increase and decrease

17 a Increase $550 by 20%

   b Decrease $3600 by $33\frac{1}{3}\%$

18 A photocopier is discounted by 35% from its selling price of $1180. Calculate its reduced price.

2-05 The unitary method

19 Selma donated $27 to charity, which was 15% of her savings. What was her total savings?

2-06 Profit, loss and discounts

20 If the selling price of a lounge suite was $900 and the cost price was $750, find:

   a the profit

   b the profit as a percentage of the cost price.

21 Add 10% GST to the cost of a digital tablet priced at $205.

ISBN 9780170351058

EARNING AND SAVING MONEY

3

IN THIS CHAPTER YOU WILL:

- calculate wages and salaries
- convert between weekly, fortnightly, monthly and annual incomes
- calculate overtime pay involving time-and-a-half and double-time
- calculate commission, piecework and annual leave loading
- calculate income tax based on allowable deductions and taxable income
- read PAYG tax tables to calculate PAYG tax
- calculate net pay from gross pay
- use the simple interest formula $I = PRN$ to calculate interest, principal and period
- solve problems involving term payments
- calculate compound interest for two to three years by repeated percentage increase
- use the compound interest formula $A = P(1 + R)^n$ to calculate final amounts and compound interest

Shutterstock.com/ pcruciatti

WORDBANK

wage An income calculated on the number of hours worked, usually paid weekly or fortnightly.

salary A fixed amount of income per year, usually paid weekly, fortnightly or monthly.

per annum / annual Per year.

1 year = 52 weeks = 26 fortnights = 12 months

EXAMPLE 1

Amy is a hairdresser who earns $32.75 per hour working Monday to Friday. Calculate Amy's wage if she works:

a 26 hours

b 30.5 hrs

SOLUTION

a Wage for 26 hours = 26 × $32.75

= $851.50

b Wage for 30.5 hours = 30.5 × $32.75

= $998.875

≈ $998.88

EXAMPLE 2

Bradley is offered a job at Costright with a starting salary of $85 000 p.a. and a job at Buyright on a weekly wage of $1586.20. Which position pays more?

✱ p.a. = per annum = per year

SOLUTION

To compare incomes, convert to weekly amounts (or yearly amounts).

Weekly pay for Costright = $85 000 ÷ 52 ⟵ 1 year = 52 weeks

= $1634.6153…

≈ $1634.62

This is more than the weekly pay at Buyright ($1586.20), so he should choose Costright.

EXAMPLE 3

Chloe is earning $4280 per month as a market researcher. Convert this to a weekly pay.

SOLUTION

As there are not an exact number of weeks in a month, we must convert the monthly pay to a yearly pay first (× 12), and then the yearly pay to a weekly pay (÷ 52)

Yearly pay = $4280 × 12 ⟵ 1 year = 12 months

= $51 360

Weekly pay = $51 360 ÷ 52 ⟵ 1 year = 52 weeks

= $987.6923…

≈ $987.69

1   Calculate Brittany's wage if she works 27.5 hours at $28.54 per hour. Select the correct answer **A**, **B**, **C** or **D**.

   A $783.75   **B** $78.49   **C** $784.85   **D** $7848.50

2   Find Samir's fortnightly wage if he works 28 hours one week and 19.4 hours the next at a rate of $25.60 per hour. Select **A**, **B**, **C** or **D**.

   A $1213.44   **B** $121.34   **C** $1256.10   **D** $524.64

3   Calculate the weekly wage for each person listed below.

   a   Jordan: 22 hours at $24.60 per hour

   b   Monique: 34 hours at $21.65 per hour

   c   Roula: 23.8 hours at $29.45 per hour

   d   Phillippe: 31.6 hours at $29.82 per hour

4   Calculate the fortnightly pay for each person in Question 3 if they work the same number of hours for both weeks.

5   For each annual salary, find:

   i   the monthly pay   ii   the fortnightly pay

   a $68 750   b $88 920   c $106 480   d $96 550

6   Calculate the annual income for each person.

   a   Iman earns a wage of $828.60 per week.

   b   Charlotte earns $4275.40 per month.

   c   Riley has a fortnightly pay of $2166.25.

7   Convert each monthly pay to a weekly pay.

   a $3275   b $5438.50

   c $4865.85   d $5946.90

8   Convert each weekly pay to a monthly pay.

   a $980   b $850.45

   c $762   d $1145.80

9   Who earns more: Lucy on $2950.25 per fortnight or Holly on a salary of $78 400 p.a.?

10 Jacinta is paid a wage of $29.50 per hour for the first 8 weeks while on probation. She works 32 hours per week. For the remainder of the year, her wage rises to $36.90 per hour. How much will she earn in her first year?

11 Convert a monthly wage of $6825 to a fortnightly pay.

WORDBANK

overtime Work done outside usual business hours and paid at a higher rate.

time-and-a-half An overtime rate which is 1.5 times the normal hourly rate.

double-time An overtime rate which is 2 times the normal hourly rate.

EXAMPLE 4

Dylan earns $28.90 per hour in IT support and works a 32-hour week from Monday to Friday. Last week, he also worked 3 hours at time-and-a-half on Saturday and 4.5 hours on Sunday at double-time. How much was his wage last week?

SOLUTION

Normal wage = $28.90 × 32

= $924.80

Saturday overtime = 3 × 1.5 × $28.90 ⟵————— Time-and-a-half (×1.5)

= $130.05

Sunday overtime = 4.5 × 2 × $28.90 ⟵————— Double-time (×2)

= $260.10

Total weekly wage = $924.80 + $130.05 + $260.10

= $1314.95

EXERCISE 3-02

1 How much is James' weekly wage if he works 30 hours at a rate of $24.90 per hour and does 4 hours overtime at time-and-a-half? Select the correct answer **A**, **B**, **C** or **D**.

   A $1269.90   **B** $896.40   **C** $846.60   **D** $859.05

2 How much is Tori's weekly wage if she works 32 hours at a rate of $26.45 per hour and does 3.5 hours overtime at double-time? Select **A**, **B**, **C** or **D**.

   A $938.98   **B** $1877.95   **C** $978.65   **D** $1031.55

3 Find the overtime pay earned for time-and-a-half working:

   a 5 hours at a normal rate of $23.50

   b 4.5 hours at a normal rate of $27.60

4 Find the overtime pay earned in Question **3** if the rate was double-time.

ISBN 9780170351058

5   Copy and complete to find the wage for a worker who earned $29.40 per hour for a 35-hour week.

Overtime was worked at time-and-a-half for 3 hours and double-time for 2.5 hours.

Normal wage = $29.40 × ___

= $ ____

Overtime = $29.40 × 1.5 × ___ + $29.40 × 2 × ____

= $_____

Total wage = $_____ + $_____

= $ _____

6   Charles works at a convenience store for 33 hours per week. If he works an extra 4 hours overtime at time-and-a-half, what is his weekly pay if his normal rate of pay is $23.50 per hour?

7   Jade works at the local fruit market for 20 hours per week at a rate of $22.60 per hour. She then works overtime for 3.5 hours on Saturday at time-and-a-half and 3 hours on Sunday at double-time. What is Jade's wage for the week?

8   Hayden works at an ice-cream shop and earns a wage of $26.85 per hour for a 26-hour week. If he works 2.5 hours overtime on Saturday at time-and-a-half and 4.5 hours overtime on Sunday at double-time, what will be his total wage?

9   Sumi works at a bowling alley and is paid $27.50 per hour on weekdays for the first 8 hours and then time-and-a-half for any time after that. On weekends she is paid double-time. Find her wage if she works 9 a.m. to 5.30 p.m. Monday to Friday and 11 a.m. to 4.30 p.m. on Saturday.

10 Liam works at the same bowling alley. Copy and complete this timesheet, then calculate his wage for the week.

Day	Start	Finish	Normal hours	Time-and-a-half	Double-time
Monday	9.00 a.m.	4.00 p.m.	7	0	0
Tuesday	8.00 a.m.	3.00 p.m.		0	
Wednesday	10.00 a.m.		8	1.5	0
Thursday	8.30 a.m.	6.30 p.m.			
Friday	8.15 a.m.		8	2.5	0
Saturday	10.00 a.m.	3.30 p.m.	0	0	
Sunday		3.20 p.m.	0		4.5

WORDBANK

commission Income for salespeople and agents, calculated as a percentage of sales.

piecework Income calculated per item made or processed, such as sewing of clothes or delivery of items.

bonus A one-off payment to employees for a job completed on time or to a high standard.

annual leave loading Also called holiday loading, this is extra pay to employees for their 4 weeks of annual holidays, calculated at 17.5% of 4 weeks pay.

EXAMPLE 5

Taylor is a real estate agent who receives a retainer of $640 per month plus 1.2% commission on her total sales for the month. Find her earnings for April if she sold a property for $560 000.

SOLUTION

Taylor's earnings = $640 + 1.2% × $560 000 ⟵——— retainer + commission

$= \$7360$

EXAMPLE 6

Nick charges $1.20 per brick for a feature wall that he is asked to build. The wall measures 3.5 m by 11 m and needs 42 bricks per square metre to complete the pattern. How much does Nick charge for the job?

SOLUTION

Calculate the number of bricks required first.

Area of the wall = 3.5 × 11 = 38.5 m²

Number of bricks = 42 × 38.5 = 1617

Job charge = 1617 × $1.20 = $1940.40

Annual leave loading = 17.5% × weekly pay × 4
Total holiday pay = 4 weeks pay + annual leave loading

EXAMPLE 7

Connor earns $1121.54 per week and takes his annual leave of 4 weeks after Christmas. Calculate his total holiday pay.

SOLUTION

Annual leave loading = 17.5% × $1121.54 × 4 ⟵——— 17.5% of 4 weeks' pay

$= \$785.078$

$\approx \$785.08$

Holiday pay = $1121.54 × 4 + $785.08 ⟵——— 4 weeks wage + annual leave loading

$= \$5271.24$

1 What is Kate's wage if she sells $3600 worth of kitchenware and earns a retainer of $420 plus a commission of 8%? Select the correct answer **A**, **B**, **C** or **D**.

   A $3300   **B** $4020   **C** $870   **D** $708

2 What are Liam's annual earnings if he is paid a salary of $108 460 p.a. and a bonus of 7.5% of his salary? Select **A**, **B**, **C** or **D**.

   A $8134.50   **B** $116 594.50   **C** $116 583.75   **D** $116 052.20

3 A real estate agent sold a new home for $620 000. His commission was 4% on the first $10 000 and 2.5% on the rest. How much was the commission?

4 Calculate each commission.

   a An insurance salesperson is paid 15% commission on $12 250 worth of premiums sold.

   b A bookseller is paid 12% commission on sales of $3890.

   c A travel agent is paid 12.5% commission on sales of $3850.

   d An actor's agent is paid a retainer of $260 per week plus 3.5% commission on $32 800 worth of income.

5 Calculate each pay earned by piecework.

   a Brendan delivers 2400 leaflets at 35c per leaflet.

   b Ania sews 240 dresses at $4.80 per dress.

   c Jackson paints 852 posts at $1.26 per post.

   d Kayla checks 1362 pages at 64c per page.

6 Calculate the annual bonus earned by each person.

   a Zoe's bonus is 7% of her $82 000 salary.

   b Matthew earned a 12% bonus of his salary, which is $3720 paid monthly.

   c Jeremy's bonus is 8.5% of his earnings, which is $690.25 per week.

   d Erin's bonus is 9.4% of her earnings, which is $1680 per fortnight.

7 Copy and complete to find the solution to this problem.

   Blake earns $865.50 per week managing a restaurant. Calculate his total holiday pay.

   4 weeks wage = $865.50 × __ = $_____

   Annual leave loading = 17.5% × $____ = $_____

   Total holiday pay = $ _____ + $ _____ = $_____

8 Calculate the annual leave loading and total holiday pay for each wage.

   a $780.60 per week

   b $1460 per fortnight

   c $5840 per month

9 Sarah earns a salary of $96 500 p.a. and decides to take 4 weeks annual leave plus an additional 3 weeks unpaid leave for a holiday to Scotland. Calculate:

   a her annual leave loading

   b her total holiday pay

   c the amount of money lost by taking 3 weeks unpaid leave.

WORDBANK

gross income The total amount earned, including bonuses and overtime.

allowable (tax) deductions Any amounts subtracted from a gross income, such as work expenses and donations to charity, that are not taxed.

taxable income Gross income less allowable deductions, the remaining income that is charged tax, rounded down to the nearest dollar.

Income tax is a tax paid to the government for services such as roads, schools and hospitals. The amount of tax paid depends upon the amount of income and is calculated according to this table.

Taxable income (rounded down to the nearest dollar)	Tax on this income
$0 - $18 200	Nil
$18 201 - $37 000	19c for each $1 over $18 200
$37 001 - $80 000	$3572 plus 32.5c for each $1 over $37 000
$80 001 - $180 000	$17 547 plus 37c for each $1 over $80 000
$180 001 and over	$54 547 plus 45c for every $1 over $180 000

Taxable income = Gross income - total allowable deductions

EXAMPLE 8

Find the tax payable on Jacob's salary of $78 400 p.a. if he has weekly deductions of $28.80 for union fees and $120 for voluntary superannuation contributions, and annual work-related expenses of $7400.

SOLUTION

Total annual deductions = $28.80 × 52 + $120 × 52 + $7400

$\qquad$ = $15 137.60

Taxable income = $78 400 − $15 137.60 $\longleftarrow$ Gross income − allowable deductions

$\qquad$ = $63 262.40

$\qquad$ = $63 262 $\longleftarrow$ Rounded down to the nearest dollar

Tax payable = $3572 + 0.325 × ($63 262 − 37 000)

$\qquad$ = $12 107.15

✱ Use line 3 of the table.
32.5c for each dollar means 32.5% or 0.325

1 What is the taxable income if the gross income is $84 500 p.a. and the total deductions are $226 per week? Select the correct answer **A, B, C** or **D**.

   A $84 274   **B** $74 724   **C** $72 748   **D** None of these

2 Calculate the tax payable on a taxable income of $35 440. Select **A, B, C** or **D**.

   A $3313.60   **B** $3275.60   **C** $6733.60   **D** None of these

3 Find the total deductions p.a. for each person.

   a Alex: travel costs $46/week, union fees $22.80/fortnight, superannuation $80/week

   b Lucia: uniforms $32/week, superannuation $126/week, accommodation $280/month

   c Jessie: social club $21/week, charities $150/month, superannuation $240/fortnight

   d Stefan: travel $154/week, union fees $26.20/month, superannuation $420/quarter

4 Find the income tax payable for each person in Question **3** if:

   a Alex earns $54 500 p.a.   **b** Lucia earns $82 400 p.a.

   c Jessie earns $162 000 p.a.   **d** Stefan earns $95 720 p.a.

5 Calculate the income tax payable for each taxable income.

   a $56 850   **b** $124 560   **c** $83 410

   d $134 700   **e** $17 500   **f** $195 740

6 Kurt earns $725.80 per week gross pay as a horse trainer.

   a What is his gross pay per year?

   b What is the tax payable on his gross pay?

   c His employer contributes 9.5% of his gross pay towards his superannuation. How much is this per annum?

Bonita Cheshier | Dreamstime.com

7 **a** Megan earns a salary of $135 000 p.a. as a chiropractor. Calculate her income tax if she has annual work expenses of $9450 and voluntary superannuation contributions of $650 per fortnight.

   b What percentage (correct to one decimal place) of Megan's gross income is income tax?

WORDBANK

PAYG tax Stands for Pay As You Go tax, which is income tax taken from your pay every payday to avoid you paying a huge amount of tax at the end of the financial year.

net pay Gross pay less tax and other deductions.

Net pay = Gross pay – tax – other deductions

EXAMPLE 9

Tuan worked at JG Hi Fi during summer break and earned $391 per week for 6 weeks.

a Use this PAYG tax table to calculate the amount of tax deducted from Tuan's pay each week.

Weekly earnings ($)	PAYG tax withheld ($)
384 – 386	87
387 – 389	88
390 – 392	89
393 – 394	90

b Calculate Tuan's net pay for the 6 weeks if he also contributed $45 per fortnight into his superannuation.

SOLUTION

a Tuan's PAYG tax = $89   ←——————— Using line 3 of the table for $391

b Tuan's gross pay = $391 × 6

$= 2346

Tuan's net pay = $2346 – 6 × $89 – 3 × $45   ←——— 6 × PAYG tax, 3 × superannuation

$– 1677

EXAMPLE 10

Nikki earns $1248 per fortnight as a ski instructor. Her fortnightly deductions are $105 for travel expenses and $52.50 for health insurance. Calculate Nikki's PAYG tax and net pay per fortnight.

Fortnightly earnings ($)	PAYG tax withheld ($)
1240 – 1244	336
1246 – 1250	338
1252 – 1256	340
1258 – 1262	342

SOLUTION

Nikki's PAYG tax = $338 per fortnight   ←——— Using line 2 of the table for $1248

Nikki's net pay = $1248 – $338 – $105 – $52.50 ←——— Gross income – deductions

$= 752.50

ISBN 9780170351058

1   Use the weekly PAYG table to find the net pay of a person earning $385 per week. Select the correct answer **A, B, C** or **D**.

   A $295   **B** $296   **C** $297   **D** $298

2   Use the fortnightly PAYG table to find the net pay of a person earning $1254 per fortnight. Select **A, B, C** or **D**.

   A $918   **B** $916   **C** $914   **D** $9912

3   Find the PAYG tax payable on each wage.

   a $392/week for 5 weeks

   b $388/week for 3.5 weeks

   c $22/hour for 17.5 hours/week

   d $24/hour for 16.2 hours/week

   e $1242 per fortnight for 4 fortnights

   f $1260 per fortnight for 3 fortnights

   g $24.20/hour for 52 hours/fortnight

   h $20.80/hour for 60 hours/fortnight

4   Ashleigh's gross pay is $680 per week. If she has deductions of $120.50 for tax and $198 for loan repayments, what is her weekly net pay?

5   Trent works a 30-hour week and is paid $22.80 per hour.

   a   What is his gross weekly wage?

   b   If taxation of 33% is deducted, what is his net pay?

6   Copy and complete the following.

Name:	Kate Keneally	Date:	16/06/16 to 20/6/16
Hourly rate:	$27.80	**Gross pay:**	
Hours worked:	33.5	**Tax:**	
Tax rate:	26.5%	**Net pay:**	

7   Pierre worked at an ice-skating rink during his holidays and earned $388 per week for 5 weeks. Calculate:

   a   his gross pay for the 5 weeks

   b   the total PAYG tax that was deducted

   c   Pierre's net pay for the 5 weeks if he contributed $50 per fortnight to his superannuation.

8   Zoe earns a gross income of $1260 per fortnight. Her deductions are $110 for superannuation and $48.80 for private health insurance, as well as PAYG tax.

   a   Calculate Zoe's net pay per fortnight.

   b   Find Zoe's deductions as a percentage of her gross pay, correct to one decimal place.

WORDBANK

interest Income earned from investing money or a charge spent for borrowing money.

principal The amount invested or borrowed. Interest is calculated on this amount.

simple interest Interest calculated on the original principal, also called flat-rate interest.

THE SIMPLE INTEREST FORMULA

Interest = principal × interest rate per period (as a decimal) × number of time periods

$I = PRN$

EXAMPLE 11

Nicola borrowed $12 500 to buy a car. She is charged simple interest by the bank at a rate of 5% p.a. for 3 years. Calculate the simple interest charged.

SOLUTION

$P = \$12\,500$, $R = 5\%$ p.a. $= 0.05$ as a decimal, $N = 3$ years

> ✳ Make sure R and N are given in the same time units; for example, both in years or both in months.

$I = PRN = \$12\,500 \times 0.05 \times 3 = \$1875$

Nicola is charged $1875 simple interest over 3 years.

EXAMPLE 12

Find the simple interest earned on $15 600 at 8.5% p.a. for 120 days.

SOLUTION

$P = \$15\,600$, $R = 8.5\%$ p.a. $= 0.085$ p.a. $= \dfrac{0.085}{365}$ per day, $N = 120$ days

> ✳ R and N are both in days.

$I = PRN = \$15\,600 \times \dfrac{0.085}{365} \times 120 = \$435.9452\ldots \approx \$435.95$

EXAMPLE 13

If the simple interest on an investment of $5000 is $1350 over 6 years, find the rate of interest p.a.

SOLUTION

$P = \$5000$, $I = \$1350$, $N = 6$ years, $R = ?$

$I = PRN$

$\$1350 = \$5000 \times R \times 6$

$\$1350 = \$30\,000R$

$\dfrac{1350}{30\,000} = R$

$R = 0.045 = 0.045 \times 100\% = 4.5\%$

1   Calculate the simple interest on $800 at 4% p.a. for 3 years. Select the correct answer **A, B, C** or **D**.

   A $120   **B** $96   **C** $64   **D** $72

2   Find the simple interest on $4500 at 7% p.a. for 4 years. Select **A, B, C** or **D**.

   A $12 600   **B** $1440   **C** $1260   **D** $1288

3   Tyler invested $4000 in a savings account paying 5% p.a. simple interest.

   a   How much interest did he receive in the first year?

   b   How much interest did he receive after 10 years?

4   Copy and complete this solution.

   To find the simple interest on $500 for 3 years at 7% p.a.:

   $P = \$500, R = \_\_\% = 0.07$ as a decimal, $N = \_\_$ years

   $I = PRN$

   $= \$500 \times \_\_\_\_ \times 3$

   $= \$\_\_\_\_$

5   Calculate the simple interest on each investment.

   a $850 for 3 years at 5% p.a.   **b** $3500 for 6 years at 7% p.a.

   c $5200 for 2 years at 5.5% p.a.   **d** $45 000 for 9 years at 3.6% p.a.

6   Danielle invested $2500 in a finance company at 6% p.a.

   a   How much interest did she receive in the first year?

   b   Find the total interest paid over a period of 7 years.

7   Copy and complete this solution.

   To find the simple interest on $6800 for 7 months at 9% p.a.:

   $P = \$6800, R = \_\_\% = 0.09$ p.a. $= \_\_\_\_$ per month, $N = \_\_$months

   $I = PRN$

   $= \$6800 \times \_\_\_\_ \times 7$

   $= \$\_\_\_\_$

8   Calculate the simple interest on each investment.

   a $720 for 3 months at 6% p.a.   **b** $8400 for 6 months at 2% per month.

   c $12 400 for 8 months at 12% p.a.   **d** $56 000 for 200 days at 7.8% p.a.

9   If the simple interest on an investment of $15 000 is $5250 at 5% p.a., how many years was the investment earning interest?

10   If the simple interest on an investment of $24 000 is $5184 over 4 years, find the interest rate p.a. as a percentage, correct to 1 decimal place.

WORDBANK

term payments A method of paying for an expensive item by making regular partial payments over a period of time, with interest being paid. Also called 'hire-purchase'.

deposit A large 'down-payment' for an expensive item that is made before the regular term payments begin.

EXAMPLE 14

A home cinema system is advertised for $1299 cash or $400 deposit followed by 24 weekly payments of $45.

a What is the total cost of purchasing the home cinema system by term payments?

b What is the interest charged if paying by term payments?

SOLUTION

a Total cost = $400 + 24 × $45   ←——— Deposit + 24 payments

          = $1480

b Interest charged = $1480 – $1299 ←——— Total cost – cash price

                = $181

EXAMPLE 15

The price of a new car was $38 000, but Jasmine bought it on terms for 15% deposit and repayments of $620 per month for 5 years.

a How much deposit did Jasmine pay?

b What was the total price paid for the car?

c How much interest did Jasmine pay for the car by buying on terms?

SOLUTION

a Deposit = 15% of $38 000

         = $5700

b Total paid = $5700 + $620 × 12 × 5 ←——— 12 months/year for 5 years

           = $42 900

c Interest = $42 900 – $38 000   ←——— Total cost – cash price

        = $4900

1   An outdoor furniture set with a marked price of $2500 can be bought with 5% deposit followed by monthly payments of $140 over 2 years. Calculate the deposit. Select the correct answer **A, B, C** or **D**.

   A $250   **B** $125   **C** $1250   **D** $12.50

2   How much would the outdoor furniture set in Question **1** cost if you are paying for it on terms? Select **A, B, C** or **D**.

   A $3485   **B** $3360   **C** $2500   **D** $2905

3   A new car costs $35 000. The dealer asks for 15% deposit plus $680 per month for 5 years.
   a   How much is the deposit?
   b   How many months are there in 5 years?
   c   How much do the payments amount to?
   d   What will be the total cost of the car on terms?

4   For each item, using term payments, calculate:
   i   the total price paid   ii   the interest paid

   a
   | Sound system |
   | Cash ... $750 or |
   | $100 deposit plus |
   | 36 payments of |
   | $21.20 |

   b
   | Television |
   | Cash ... $1250 or |
   | $275 deposit plus |
   | 24 payments of |
   | $45.70 |

   c
   | Bike |
   | Cash ... $419 or |
   | $100 deposit plus |
   | $32.50 per month |
   | for 1 year |

   d
   | Second-hand car |
   | Cash ... $5000 or |
   | 25% deposit |
   | plus 36 payments |
   | of $111.75 |

5   The cash price of a new ride-on mower was $2200. On terms, I had to pay a deposit of $750 followed by monthly payments of $82.50 for 2 years.
   a   How much was the deposit?
   b   How many monthly payments were made?
   c   What was the total of the monthly payments?
   d   What was the total cost paid under terms?
   e   How much interest was charged?

6   What is the price paid on terms for a set of golf clubs with a cash price of $1500, deposit of 10%, plus monthly payments of $50 over four years?

7   The cash price of a sound system was $550. To buy it on terms, 20% of the cash amount was needed for the deposit, plus 12 payments of $58.50.
   a   How much was the deposit?
   b   What do the payments amount to?
   c   What is the total price paid on terms?

Compound interest

While **simple interest** is calculated on the original principal only, **compound interest** is calculated on the principal plus any interest added previously. The interest earned is added to the principal so that next time, the interest is calculated on a larger principal. With compound interest, more interest is earned because we are 'earning interest on the interest'.

EXAMPLE 16

Lisa invests $4000 in an investment account for 3 years, where the interest is 6% p.a. compounded annually.

a Find the value of her investment after 3 years.

b Calculate the compound interest earned over the 3 years.

 'Compounded annually' means that the interest is calculated at the end of each year.

SOLUTION

a Interest after the **1st year** = 6% × $4000 = $240

New principal = $4000 + $240 = $4240

Interest after the **2nd year** = 6% × $4240 = $254.40

New principal = $4240 + $254.40 = $4494.40

Interest after the **3rd year** = 6% × $4494.40 = $269.66

New principal = $4494.40 + $269.66 = $4764.06

The value of Lisa's investment after 3 years is $4764.06.

b Compound interest earned = $4764.06 – $4000 ⟵————— Final amount – original principal

= $764.06

Compound interest = final amount – original principal

Compound interest involves repeated percentage increases
In Example **16**, note that adding 6% interest to the principal is the same as increasing the principal by 6% each time. We can use this fact to calculate answers more quickly.

EXAMPLE 17

A principal of $600 is invested at 10% p.a. compounded yearly. Calculate its value after two years and the total interest earned.

SOLUTION

Principal after the **1st year** = $600 + 10% × $600 = $660

Principal after the **2nd year** = $660 + 10% × $660 = $726

The value of the investment after 2 years is $726.

Compound interest earned = $726 – $600 ⟵————— Final amount – original principal

= $126

1   Increase $3000 by 8%. Select the correct answer **A**, **B**, **C** or **D**.

    A $5400   **B** $3120   **C** $3240   **D** $3024

2   A principal of $3000 is invested at 8% p.a. interest compounded annually. What is its value after 2 years? Select **A**, **B**, **C** or **D**.

    A $3480   **B** $3499.20   **C** $3259.20   **D** $3240

3   Copy and complete the table below that calculates the compound interest and the final amount on an investment of $3000 for 4 years at 5% p.a.

Principal	Interest	Principal + interest
$3000	5% × $3000 = $150	$3000 + $150 = $3150
$3150	5% × $3150 = $____	$3150 + $____ = $ ____
$		

4   Copy and complete this table.

Year	Principal	Interest	Principal + interest
1	$6000	8% × $6000 = $480	$6000 + $480 = $6480
2	$6480	8% × $6480 = $____	$6480 + $____ = $ ____
3			

What is this table calculating?

5   Find the final amount of each investment and the compound interest earned.

    a $5000 invested at 3% p.a. for 2 years

    b $10 000 invested at 6% p.a. for 3 years

    c $12 600 invested at 4.5% p.a. for 1 year.

    d $4000 invested at 10% p.a. for 2 years

6   Sofija saves $12 000 towards a North American holiday and then invests it at 8% p.a. interest for 2 years. Calculate how much money Sofija will have for her holiday.

Compound interest formula

THE COMPOUND INTEREST FORMULA

$A = P(1 + R)^n$, where:

   A = final amount
   P = principal (original amount)
   R = interest rate per period (as a decimal)
   n = number of compounding periods

Compound interest = final amount − principal

EXAMPLE 18

Lisa puts $4000 in an investment account for 3 years, where the interest is 6% p.a. compounded annually.

a Find the value of her investment after 3 years.

b Calculate the compound interest earned over the 3 years.

SOLUTION

This is the same problem from Example 16 on page 52, but solved using the compound interest formula.

a $P = \$4000$, $R = 6\%$ p.a. $= 0.06$, $n = 3$ years

✱ Make sure R and n are given in the same time units, for example, both in years or both in months.

$$A = P(1 + R)^n$$
$$= \$4000(1 + 0.06)^3$$
$$= \$4000(1.06)^3 \quad \longleftarrow \quad \text{On a calculator: } 4000 \; \boxed{\times} \; 1.06 \; \boxed{x^y} \; 3 \; \boxed{=}$$
$$= \$4764.064$$
$$\approx \$4764.06$$

b Compound interest $= \$4764.06 - \$4000 \quad \longleftarrow \quad$ final amount − principal
$$= \$764.06$$

EXAMPLE 19

Calculate the final amount if $4000 is invested at 6% p.a. for 3 years and interest is compounded monthly.

SOLUTION

$P = \$4000$, $R = 6\%$ p.a. $= 0.06$ p.a. $= \dfrac{0.06}{12}$ per month $= 0.005$ per month, $n = 3$ years $= 36$ months

✱ R and n are both in months.

$$A = P(1 + R)^n$$
$$= \$4000(1 + 0.005)^{36}$$
$$= \$4000(1.005)^{36} \quad \longleftarrow \quad \text{On a calculator: } 4000 \; \boxed{\times} \; 1.005 \; \boxed{x^y} \; 36 \; \boxed{=}$$
$$= \$4786.7220\ldots$$
$$\approx \$4786.72$$

ISBN 9780170351058

1   Copy and complete this table.

Principal	Interest rate	Time	Final amount $A = P(1 + R)^n$	Compound interest $A - P$
$2000	8% p.a.	3 years		
$5800	7% p.a.	6 years		
$9400	5% p.a.	3.5 years		
$12 700	4.5% p.a.	8 years		
$25 000	6% p.a. compounded monthly	8 months		

2   If $5000 is invested for 3 years at 9% p.a. interest compounded monthly, what are the values of R and n in the compound interest formula? Select the correct answer **A**, **B**, **C** or **D**.

   A $R = 0.09, n = 3$ **B** $R = 0.09, n = 36$
   C $R = 0.0075, n = 36$ **D** $R = 0.0075, n = 3$

3   What is the final amount of $5000 in Question **2**? Select **A**, **B**, **C** or **D**.

   A $6543.23   **B** $6475.15
   C $5113.35   **D** $111 256.13

4   Calculate the final amount and compound interest earned with each investment.

   a $4000 at 5% p.a. over 3 years, interest compounded annually.
   b $5500 at 6% p.a. over 4 years, interest compounded monthly.
   c $10 600 at 8% p.a. over 5 years, interest compounded quarterly.
   d $16 200 at 9% p.a. over 4.5 years, interest compounded monthly.
   e $64 000 at 4% p.a. over 7 years, interest compounded quarterly.

5   Calculate the compound interest earned on an investment of $100 000 at 6% p.a. over 25 years, where interest is compounded.

   a   annually   **b**   monthly

6   Joel works part time and earns $380 per weekend. He is able to save 80% of his wage each week. He works for 16 weekends and then deposits his savings in an account that earns 5% p.a. interest compounded annually, for 4 years. Will he have enough to go on a European cruise costing $5800?

CLUELESS CROSSWORD

Make a copy of this puzzle, then complete the crossword using words from this chapter.

C	O		M		S		I		N			T					
M				P		E		E	W		R						
P		Y															
			G	R			S		G		S				P		
U												L					
N		T		N				D		P		S		T			
		C		N	S	U		E				A		N			
I								H		D		C					
			S		L		R	Y									
T		M		T				L				P					
					E	A		N		G							
R				O				D				L					
		B		N		S		R		T	E						
S								Y									
T		X		T		O											

Part A General topics

Calculators are not allowed.

1 Convert $\dfrac{36}{8}$ to a mixed numeral.

2 Evaluate 8.4 – 4.25.

3 Arrange these integers in descending order: –5, 7, 0, 6, 3, –2

4 Simplify $xy - px + 5xy + xp$.

5 Find the mean of: 8, 3, 2, 6, 5, 5, 6.

6 Find x.

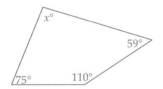

7 Convert 3:30 p.m. to 24-hour time.

8 Find the value of d if the area of this rectangle is 144 m².

9 Paula and Dane share ownership of a shop in the ratio 5 : 7. If the shop makes a profit of $18 900, what is Dane's share?

10 A trolley bag with a marked price of $84 is discounted by 10%. Calculate its sale price.

Part B Earning and saving money

Calculators are allowed.

3–01 Wages and salaries

11 Find the weekly wage of Brooke who earns $22.90 per hour for 31 hours' work. Select the correct answer **A, B, C** or **D**.

   A $70.99   **B** $7099   **C** $706.80   **D** $709.90

12 Find the fortnightly pay of Mikhail who earns $92 600 p.a. Select **A, B, C** or **D**.

   A $1780.77   **B** $3561.54   **C** $7716.67   **D** $3704

3–02 Overtime pay

13 Find Rebecca's weekly pay if she works 28 hours at $25.60 per hour and 4 hours overtime at time-and-a-half. Select **A, B, C** or **D**.

   A $819.20   **B** $921.60   **C** $870.40   **D** $1228.80

14 Find Scott's weekly pay if he works 30 hours at $27.40 per hour and 3 hours overtime at double-time.

3–03 Commission, piecework and leave loading

15 Georgia sells real estate and is paid commission of 4% on the first $10 000 and 2.5% on the remaining selling price of each home sold. Find her commission on a house that sold for $760 000.

16 Jeremy earns $680.60 per week at his local supermarket. He takes 4 weeks annual leave and is paid his normal wage for 4 weeks plus 17.5% annual leave loading. Calculate Jeremy's:

a annual leave loading.

b holiday pay.

3–04 Income tax

17 Use the income tax table on page 44 to calculate the income tax payable for each person.

a Brendan: Salary $72 500 p.a., deductions $165/week

b Lauren: Wage $855/week, deductions $128.40/week

3–05 PAYG tax and net pay

18 Use the PAYG tax tables on page 46 to calculate the PAYG tax payable for each person.

a Emily earns $388 per week for 6 weeks

b Jason earns $1254 per fortnight for 12 weeks

3–06 Simple interest

19 Find the simple interest on each investment.

a $2500 at 4% p.a. for 3 years

b $9500 at 6% p.a. for 7 months

3–07 Term payments

20 What will be the total cost of a car with a marked price of $32 000 if it is bought on terms for 15% deposit with monthly repayments of $685 for 4 years?

3–08 Compound interest

21 Find the final amount that $50 000 grows to at 4.5% p.a. interest compounded annually after 3 years.

3–09 Compound interest formula

22 a Find the amount that $68 000 grows to over 4 years at 6% p.a. interest compounded monthly.

b What is the compound interest earned on this amount?

ALGEBRA

4

IN THIS CHAPTER YOU WILL:

- convert worded descriptions into algebraic expressions
- substitute into algebraic expressions
- add, subtract, multiply and divide algebraic terms
- expand algebraic expressions
- factorise algebraic terms and expressions

Shutterstock.com/danielo

ISBN 9780170351058

WORDBANK

pronumeral or variable A letter of the alphabet such as a, b, c, x or y that represents a number.

algebraic expression A relationship of pronumerals, numbers and operations written in algebraic form. For example, $2x - 3$, $5a + 2b$, $x^2 - 3x$

Mathematical word	Meaning
sum, total, increase or plus	add (+)
difference or decrease	subtract (−)
product	multiply (×)
quotient	divide (÷)
twice or double	multiply by 2 (× 2)
triple	multiply by three (× 3)
square	multiply by itself (x^2)

EXAMPLE 1

Write each statement as an algebraic expression.

a Triple a

b The product of x and y

c Decrease m by 6

d The quotient of a and b

e Twice the sum of m and n

f w squared

SOLUTION

a $3a$

b xy

c $m - 6$

d $\dfrac{a}{b}$

e $2(m + n)$

f w^2

EXAMPLE 2

a Find an algebraic expression for the perimeter of the rectangle.

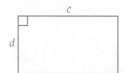

b Farmer Bob has m chickens. Write an expression for the total number of legs on the chickens.

SOLUTION

a Perimeter of the rectangle $= 2 \times c + 2 \times d$
$$= 2c + 2d$$

b Number of legs $= 2 \times m$
$$= 2m$$

iStockphoto.com/Granicapa

1 To find triple the sum of two numbers, which is the correct order? Select the correct answer **A**, **B**, **C** or **D**.

   A multiply by 3 and then add   **B** divide by 3 and then add

   C add and then multiply by 3   **D** add and then divide by 3

2 How many minutes are there in x hours? Select **A**, **B**, **C** or **D**.

   A $x + 60$ **B** $\dfrac{x}{60}$ **C** 60   **D** $60x$

3 Write the algebraic expression for:

   a the difference between x and y **b** the product of x and y

   c the quotient of x and y **d** the sum of x and y

4 Is each statement true or false?

   a The product of 6, a, b and d is $6abd$

   b The sum of m, 3, n and 8 is $m + n + 11$

   c The difference between $4a$ and $9b$ is $9a - 4b$

   d The quotient of $12x$ and $5y$ is $\dfrac{12x}{5y}$

5 Write each statement as an algebraic expression.

   a twice a plus 4   **b** 16 decreased by triple b

   c double the difference of m and n **d** triple the sum of x and y

   e twice n minus 8   **f** m squared less 3

   g triple y plus x **h** double the product of x and y

   i the square of the sum of m and n **j** 12 plus g squared

   k 180 plus twice d **l** m squared minus double n

   m 15 less triple b **n** decrease h by twice j

6 Write an algebraic expression for the cost of:

   a 6 bananas at a cents each   **b** 4 pizzas at \$$q$ each

   c d hours work at \$18/h   **d** 1 share when 5 winners win \$$k$

   e k pencils at \$$w$ each   **f** 1 bag if 8 bags cost \$$d$

7 For each figure below, write an algebraic expression for the perimeter.

   a   **b**   **c**

8 Write down an algebraic expression for the area of each figure in Question 7.

9 If Eva is w years old and Jake is triple her age, find in terms of w:

   a Jake's age   **b** Eva's age 3 years ago

   c Jake's age in 8 years time   **d** Eva's age in 6 years time

10 Old McDonald had a farm with x sheep and y turkeys. Find the total number of :

   a sheep legs   **b** turkey legs   **c** legs   **d** heads

WORDBANK

substitution Replacing a variable with a given value in an algebraic expression to find the value of the expression.

formula An algebraic rule using variables and an equals sign, such as $A = \frac{1}{2}bh$ for the area of a triangle.

EXAMPLE 3

If $a = 6$ and $b = -2$, evaluate each algebraic expression.

a $4a - 2b$ **b** $8ab$ **c** $4b^2 - 3a$

SOLUTION

a $4a - 2b = 4 \times 6 - 2 \times (-2)$
            $= 28$

b $8ab = 8 \times 6 \times (-2)$
        $= -96$

c $4b^2 - 3a = 4 \times (-2)^2 - 3 \times 6$
            $= -2$

EXAMPLE 4

The formula $A = \frac{1}{2}bh$ gives the area of a triangle with a base b and a height h.

Find the area of the triangle using this formula.

6.2 m

5.4 m

SOLUTION

$A = \frac{1}{2}bh$

$= \frac{1}{2} \times 5.4 \times 6.2$ ⟵   base $= b = 5.4$ and height $= h = 6.2$

$= 16.74 \text{ m}^2$

EXAMPLE 5

Complete this table of values using the formula $y = 3x - 2$

x	-1	0	1	2
y				

SOLUTION

$y = 3x - 2$

x	-1	0	1	2
y	-5	-2	1	4

     $3 \times (-1) - 2$ $3 \times 0 - 2$ $3 \times 1 - 2$ $3 \times 2 - 2$

1   Evaluate $8cd$ if $c = 5$ and $d = -3$. Select the correct answer **A**, **B**, **C** or **D**.

   A 96   **B** −120   **C** −64   **D** 120

2   Evaluate $12a^2$ if $a = -4$. Select **A**, **B**, **C** or **D**.

   A 192   **B** −48   **C** 48   **D** −192

3   If $x = 4$ and $y = -3$, is each equation true or false?

   a $x + y = -1$ **b** $2x - y = 11$ **c** $3xy = 36$ **d** $\dfrac{6x}{y} = -6$

4   Find the value of each algebraic expression if $a = -4$ and $b = -2$.

   a $3ab$ **b** $4a - 2b$ **c** ab^2 **d** $5b - 3a$

   e $4(2a + b)$ **f** $8b^2$ **g** $16b - a$ **h** $3ab^2$

5   Copy and complete this table.

	$4x - y$	$3x^2$	$\dfrac{4x}{y}$	$2(3x - y)$	$4xy^2$
$x = 2, y = 4$					
$x = -1, y = 3$					
$x = 4, y = -2$					
$x = -1, y = -3$					
$x = 5, y = -2$					

6   Evaluate each formula.

   a $P = 2l + 2w$ where $l = 2.4$ and $w = 1.8$ **b** $A = \dfrac{1}{2}bh$ where $b = 16$ and $h = 7$

   c $A = lw$ where $l = 6.8$ and $w = 3.7$ **d** $V = lwh$ where $l = 10.3$, $w = 4$ and $h = 1.6$

   e $c = \sqrt{a^2 + b^2}$ where $a = 7$ and $b = 24$ **f** $f = \dfrac{uv}{u + v}$ where $u = 3$ and $v = 6$

7   Copy and complete each table of values.

   a $y = 2x + 1$

x	−1	0	2	3
y				

   b $y = 6 - x$

x	−1	0	1	2
y				

   c $y = 3x^2$

x	−1	0	2	3
y				

   d $y = \dfrac{3x}{2} - 2$

x	−2	0	2	4
y				

8   If $V = \pi r^2 h$ is the formula for the volume of a cylinder, find, correct to one decimal place, the volume of a cylinder with radius 5 cm and height 4.8 cm.

WORDBANK

like terms Terms with exactly the same variables, for example, $2a$ and $3a$, $7y$ and $-3y$, $4ab$ and $-2ba$.

- ■ Only like terms can be added and subtracted.
- ■ The sign in front of a term belongs to it.
- ■ x means $1x$, the '1' does not need to be written down.

EXAMPLE 6

Simplify each expression

a $8a - 3a$ **b** $-2y - y$ **c** $9x^2 - 3x + 2x^2$ **d** $8a - 6b + a - 5b$

SOLUTION

a $8a - 3a = 5a$ **b** $-2y - y = -3y$

Group together like terms, including the sign in front of it.

c $\underline{9x^2} - 3x \underline{+ 2x^2} = 11x^2 - 3x$ **d** $\underline{8a} - 6b \underline{+ a} - 5b = 9a - 6b - 5b$
$$= 9a - 11b$$

EXAMPLE 7

Write an algebraic expression for the perimeter of this rectangle.

SOLUTION

Perimeter $= 4a + 5 + 2a + 4a + 5 + 2a$ ←—— Perimeter is the sum of the four sides.
$= 4a + 2a + 4a + 2a + 5 + 5$ ←—— Group like terms
$= 12a + 10$

iStockphoto.com/Baloncici

1. Which are the like terms in the expression $3a - 4b + a - 3d$? Select the correct answer **A**, **B**, **C** or **D**.

   A $3a, -3d$ **B** $3a, a$ **C** $-4b, -3d$ **D** $3a, -4b$

2. Simplify $5x - 4y - y + 5$. Select **A**, **B**, **C** or **D**.

   A $10x - 5y$ **B** $5x - 4y + 5$ **C** $5x - 5y + 5$ **D** $10x - 5y + 5$

3. In each list, write down the like terms.

   a $2x, 2y, 4x, 2$ b $3a, a, 6b, 6$

   c $4w, 2v, -2w, 4$ d $7ab, 4ba, -7a, -7b$

   e $5a^2, 3ab, 3b^2, 5ba$ f $-4x, 8y, -4xy, 12x$

4. Is each statement true or false?

   a $4x - 5x = x$ b $5y + (-2y) = 7y$

   c $12a + 8 - 4a = 8 + 8a$ d $6ab - ab = 5ba$

   e $15x^2 - 8xy = 7xy$ f $-4mn + 6m^2 - mn = -5mn + 6m^2$

5. Simplify each algebraic expression.

   a $4a - 3a$ b $5x + 3x$ c $11w - 2w - 3$

   d $15m - 5m$ e $-3y + 9y$ f $16n - 14n + 9$

   g $4a - 9a$ h $9m + 4m - m$ i $24t - 20t + 3t$

6. Copy and complete the following.

   a $3m - 2n - 8m + n = 3m - \_\_\_ - 2n + \_\_\_$ b $9x + 4y - 2y - x = 9x - \_\_\_ + 4y - \_\_\_$

                $= -5m - \_\_\_$ $= 8x + \_\_\_$

7. Simplify each algebraic expression.

   a $3x - y + 6x$ b $-4y - 2x + 7x$

   c $5m - 2n + 8m - n$ d $7m - n + 6m$

   e $5x - 3y - x + 6y$ f $16a - 5b - b - 3a$

   g $22r - 3s + r - 6s$ h $3ab - ba + 7 - 2ab$

   i $4xy - 7y - 3yx + 9$

8. Explain in words why $5ab - 2b + 4ab^2$ cannot be simplified.

9. Write an algebraic expression for the cost of

   a   6 bread rolls at x cents each

   b y bread rolls at r cents each.

10. Find a simplified algebraic expression for the perimeter of each shape.

   a   b   c

     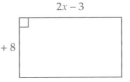

To multiply or divide algebraic terms:
- ▪ multiply or divide the numbers first
- ▪ then multiply or divide the variables
- ▪ write the variables in alphabetical order
- ▪ if dividing, write the answer in fraction form

← They do not have to be like terms!

EXAMPLE 8

Simplify each expression.

a $5 \times 2x$

b $3y \times 4a$

c $-5x \times 7xy$

d $-4ab \times (-2b)$

e $12c \div 4$

f $35ab \div 5a$

g $36mn^2 \div (-9n)$

h $-24xy \div (-6yz)$

SOLUTION

a $5 \times 2x = 5 \times 2 \times x$
$$= 10x$$

b $3y \times 4a = 3 \times 4 \times y \times a$
$$= 12ay$$

c $-5x \times 7xy = -5 \times 7 \times x \times xy$
$$= -35x^2y$$

d $-4ab \times (-2b) = -4 \times (-2) \times ab \times b$
$$= 8ab^2$$

e $12c \div 4 = \dfrac{12c}{4}$
$$= 3c$$

f $35ab \div 5a = \dfrac{35ab}{5a}$
$$= 7b$$

g $36mn^2 \div (-9n) = \dfrac{36mn^2}{-9n}$
$$= -4mn$$

h $-24xy \div (-6yz) = \dfrac{-24xy}{-6yz}$
$$= \dfrac{4x}{z}$$

EXAMPLE 9

Find an algebraic expression for the area of each shape.

a

b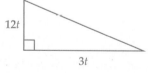

SOLUTION

a Area of a rectangle = $8x \times 7y$
$$= 56xy$$

b Area of triangle = $\dfrac{1}{2} \times 3t \times 12t$
$$= 18t^2$$

ISBN 9780170351058

EXERCISE 4–04

1   Simplify $-5x \times (-3xy)$. Select the correct answer **A**, **B**, **C** or **D**.

    A $-15xy$ **B** $15x^2y$ **C** $-15x^2y$ **D** $15xy^2$

2   Simplify $-16ab \div 8b$. Select **A**, **B**, **C** or **D**.

    A $-8a$ **B** $2a$ **C** $-2ab$ **D** $-2a$

3   Simplify each algebraic expression.

    a $4 \times 6m$ **b** $-2x \times 4y$

    c $-8m \times 3n$ **d** $9ab \times (-4b)$

    e $6bc \times (-3cd)$ **f** $8a^2 \times 3a$

    g $12m \times 3m^3$ **h** $-4w^2 \times (-3uw)$

    i $-8xy \times (-4y^2)$ **j** $7ab^2 \times (-3ab)$

    k $16x \div 2$ **l** $18b \div (-3)$

    m $\dfrac{-15d}{-3}$ **n** $25ab \div (-5)$

    o $\dfrac{32bc}{-8}$ **p** $75xy \div (-25)$

    q $\dfrac{-54mn}{9n}$ **r** $-36ab \div (-3bc)$

    s $18vw \div (-6wx)$ **t** $27r^2 \div (-9r)$

    u $\dfrac{125xy}{-25y^2}$

4   Find a simplified algebraic expression for the area of each shape.

    a 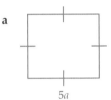   **b** 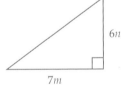   **c**

5   Simplify each expression.

    a $\dfrac{16ab^3}{8b}$ **b** $\dfrac{25x^3y^2}{5xy}$

    c $\dfrac{-40m^3n}{8mn^3}$ **d** $12 - 3x \times 4$

    e $6 + 14a \div 7$ **f** $-24ab \div (-6b) \times 2a$

6   If the area of a rectangle is $48np$ and its length is $8p$, what is its width?

7   If the area of a triangle is $32x^2$ and its height is $2x$, what is the length of its base?

WORDBANK

expand To remove the brackets or grouping symbols in an algebraic expression.

When multiplying 14 by 9 mentally, we know that

$14 \times 9 = 14 \times (10 - 1)$
$= 14 \times 10 - 14 \times 1$
$= 140 - 14$
$= 126$

This idea can be used to expand algebraic expressions.

> **To expand an algebraic expression with brackets**, the term outside the brackets must be multiplied by every term inside the brackets.
> $a(b + c) = ab + ac$

EXAMPLE 10

Expand each algebraic expression.

a $3(x + 5)$ **b** $4(3a - 2)$ **c** $-6(2x - 4)$ **d** $a(3a + 5)$

SOLUTION

a $3(x + 5) = 3 \times x + 3 \times 5$
$= 3x + 15$

b $4(3a - 2) = 4 \times 3a + 4 \times (-2)$
$= 12a - 8$

c $-6(2x - 4) = -6 \times 2x + (-6) \times (-4)$
$= -12x + 24$

d $a(3a + 5) = a \times 3a + a \times 5$
$= 3a^2 + 5a$

EXAMPLE 11

Expand and simplify each expression.

a $5(2x - 4) - 3(x + 2)$ **b** $a(2a - 6) - (3a - 4)$

SOLUTION

a $5(2x - 4) - 3(x + 2) = 5 \times 2x + 5 \times (-4) + (-3) \times x + (-3) \times 2$ ⟵   Expand brackets first
$= 10x - 20 - 3x - 6$
$= 7x - 26$

b $a(2a - 6) - (3a - 4) = a \times 2a + a \times (-6) + (-1) \times 3a + (-1) \times (-4)$ ⟵ $-(3a - 4)$ means $-1(3a - 4)$
$= 2a^2 - 6a - 3a + 4$
$= 2a^2 - 9a + 4$

ISBN 9780170351058

1   Expand $4(3a - 5)$. Select the correct answer **A, B, C** or **D.**

   A $12a - 5$ **B** $7a - 20$ **C** $12a - 20$ **D** $7a - 5$

2   Expand $3x(2x - 3)$. Select **A, B, C** or **D.**

   A $6x - 9$ **B** $6x^2 - 9$ **C** $6x^2 - 9x$ **D** $5x^2 - 6x$

3   Copy and complete each expansion.

   a $2(3x - 4) = 2 \times 3x + 2 \times$ ____   **b** $-5(3a - 1) = -5 \times 3a + (-5) \times$ ____

                $= 6x +$ ____ $= -15a +$ ____

   c $m(3m + 4) = m \times 3m + m \times$ ____   **d** $-x(2x - 4) = -x \times 2x + (-x) \times$ ____

                $= 3m^2 +$ ____ $= -2x^2 +$ ____

4   Expand each expression.

   a $4(x + 2)$ **b** $2(a - 3)$ **c** $5(2a - 1)$ **d** $6(3x + 1)$

   e $8(4 - 2a)$ **f** $a(4a - 3)$ **g** $x(2x + 3)$ **h** $2m(m - 7)$

   i $4w(2w + 6)$ **j** $2a(3a - 5)$ **k** $-3(4a + 2)$ **l** $-4(6 - 3m)$

   m $-(3x - 8)$ **n** $-(2a + 9)$ **o** $-7(2w - 5)$

5   **a**   Check that $4(x + 2) = 4x + 8$ by substituting $x = 5$ into both sides of the equation and testing whether the values are equal.

   b   Substitute another value of x into both sides and check whether the values are still equal.

6   Copy and complete each expansion.

   a $4(2y - 1) + 3y = 8y -$ ____ $+$ ____

                $=$ ____ $-$ ____

   b $16 + 2(3m - 1) = 16 +$ ____ $-$ ____

                $=$ ____ $+$ ____

   c $3(3x - 2) - 2(x + 4) = 9x -$ ____ $- 2x -$ ____

                  $=$ ____ $-$ ____

   d $4(2a - 1) + 2(3a + 1) = 8a -$ ____ $+$ ____ $+ 2$

                  $=$ ____ $-$ ____

7   Expand and simplify each expression.

   a $4(x - 4) + 2(x - 5)$ **b** $3(x + 3) - 2(x + 5)$

   c $5(a + 6) - 3(a - 8)$ **d** $5(3x - 8) - 3(x - 4)$

   e $6(2x + 1) - 3(x + 7)$ **f** $3(2a + 9) - 2(2a - 7)$

   g $-4(2x + 3) - 2(3x - 4)$ **h** $3(5x + 1) - 4(x + 8)$

   i $-3(3a + 6) - 2(2a - 5)$

8   Expand and simplify $12(3w - 4v) - 6(4v + 5w)$.

WORDBANK

highest common factor (HCF) The largest number or algebraic term that divides into two or more numbers or algebraic terms evenly.

factorise To insert brackets or grouping symbols in an algebraic expression by taking out the highest common factor (HCF); factorising is the opposite of expanding.

$$\xrightarrow{\text{expanding}}$$
$$4(3x - 1) = 12x - 4$$
$$\xleftarrow{\text{factorising}}$$

To find the highest common factor (HCF) of algebraic terms:
- find the HCF of the numbers
- find the HCF of the variables
- multiply the HCFs together

EXAMPLE 12

Find the highest common factor of:

a $10a$ and $15abc$

b $16xy$ and $24xy^2$

SOLUTION

a   The HCF of 10 and 15 is 5.
The HCF of a and abc is a.
The HCF of $10a$ and $15abc$ is $5 \times a = 5a$

b   The HCF of 16 and 24 is 8.
The HCF of xy and xy^2 is xy.
The HCF of $16xy$ and $24xy^2$ is $8 \times xy = 8xy$

To factorise an algebraic expression:
- find the HCF of all the terms and write it in front of the brackets
- divide each term by the HCF and write the answers inside the brackets
  $ab + ac = a(b + c)$
  To check if the answer is correct, expand it.

EXAMPLE 13

Factorise each expression.

a $5a - 35$ b $18ab + 27bc$ c $6m^2 + 18mn$ d $45xy^2 - 25x^2y$

SOLUTION

a $5a - 35 = 5(a - 7)$ $\longleftarrow$ HCF = 5   b $18ab + 27bc = 9b(2a + 3c)$ $\longleftarrow$ HCF = 9b
Check each answer by expanding.

$5(a - 7) = 5a - 35$ $9b(2a + 3c) = 18ab + 27bc$

c $6m^2 + 18mn = 6m(m + 3n)$ $\longleftarrow$ HCF = 6m   d $45xy^2 - 25x^2y = 5xy(9y - 5x)$ $\longleftarrow$ HCF = 5xy
Check each answer by expanding.

$6m(m + 3n) = 6m^2 + 18mn$ $5xy(9y - 5x) = 45xy^2 - 25x^2y$

1 Factorise $24a - 16$. Select the correct answer **A**, **B**, **C** or **D**.

   A $2(12a - 8)$ **B** $4(6a - 4)$ **C** $8(3a - 3)$ **D** $8(3a - 2)$

2 Factorise $36xy + 24x^2$. Select **A**, **B**, **C** or **D**.

   A $4(9xy + 6x^2)$ **B** $2x(18y + 12x)$

   C $12x(3y + 2x)$ **D** $4x(9y + 6x)$

3 Find the HCF of each pair of terms.

   a $6a$ and $8a$ b $4m$ and $8n$

   c $18b$ and 6 d $9u$ and $18v$

   e $24m$ and $8mn$ f $16a$ and $4ab$

   g $12w$ and $16w^2$ h $18m^2$ and $24mn$

4 Copy and complete each factorisation.

   a $5a + 5b = 5(\underline{\quad})$ b $3x - 6y = 3(\underline{\quad})$

   c $6m + 9n = 3(\underline{\quad})$ d $8b^2 - 6 = 2(\underline{\quad})$

   e $9a + 3b = \underline{\quad}(3a + b)$ f $24x - 12y = \underline{\quad}(2x - \underline{\quad})$

   g $4u + 16v = \underline{\quad}(u + \underline{\quad})$ h $9n^2 - 3 = \underline{\quad}(\underline{\quad} - 1)$

5 Factorise each expression, and check your answer by expanding.

   a $3a + 12$ b $9p - 18$ c $4n + 28$ d $6y - 30$

   e $9a + 54$ f $15x - 45$ g $12a + 72$ h $16c - 32$

   i $25a + 75$ j $55x - 5y$ k $a^2 + am$ l $ab + bc$

   m $mn + m^2$ n $4p - 24q$ o $xyz - 6xy$

6 Copy and complete each factorisation.

   a $2a^2 + 6a = 2a(\underline{\quad})$ b $25y - 10y^2 = 5y (\underline{\quad})$

   c $18mno + 27noq = 9no(\underline{\quad})$ d $42b^2 - 14bc = \underline{\quad}(3b - \underline{\quad})$

   e $28ab + 14bc = \underline{\quad}(2a + \underline{\quad})$ f $45x^2 - 15xy = \underline{\quad}(3x - \underline{\quad})$

7 Factorise each expression.

   a $3ap - 15pq$ b $16rs + 24st$

   c $25y - 15y^2$ d $32abc - 16bcd$

   e $4m^2 + 48m$ f $22b^2 - 11b$

   g $38fg + 19g^2$ h $3xyz - 9x$

   i $13a^2b - 26b^2$ j $36w^2 + 9vw$

   k $24r^2t - 16rt$ l $25mn^2 + 50nm^2$

8 Factorise $64a^2bc - 16b^2c + 32abc$, then expand your answer to check that it is correct.

Factorising with negative terms

To factorise an algebraic expression with a negative first term:
- include the negative sign when finding the HCF and write it in front of the brackets
- divide each term by the HCF and write the answers inside the brackets

To check if the answer is correct, expand it.

EXAMPLE 14

Factorise each expression.

a $-16x - 48$ b $-9mn - 36m$ c $-24ab + 16bc^2$

SOLUTION

a $-16x - 48 = -8(2x + 6)$ $\longleftarrow$ HCF = -8
    Check each answer by expanding.
    $-8(2x + 6) = -16x - 48$

b $-9mn - 36m = -9m(n + 4)$ $\longleftarrow$ HCF = $-9m$
    Check each answer by expanding.
    $-9m(n + 4) = -9mn - 36m$

c $-24ab + 16bc^2 = -8b(3a - 2c^2)$ $\longleftarrow$ HCF = $-8b$
    Check each answer by expanding.
    $-8b(3a - 2c^2) = -24ab + 16bc^2$

✱ Note that the signs in each bracket are different from the sign in the question as you are dividing each term by a negative number.

1 Factorise $-18x - 45$. Select the correct answer **A**, **B**, **C** or **D**.

   A $3(6x - 15)$ **B** $-9(2x + 5)$

   C $-3(6x + 15)$ **D** $-9(2x - 5)$

2 Factorise $-17mn + 34m^2$. Select **A**, **B**, **C** or **D**.

   A $-17(mn + 2m^2)$ **B** $-17m(n - 2m)$

   C $17m\,(n - 2m)$ **D** $-17m(n + 2m)$

3 Find the negative HCF of each pair of terms.

   a -3 and $-9a$ **b** -8 and $-12x$

   c $-5m$ and $-15n$ **d** $-25w$ and -30

   e $-25b$ and $-10a$ **f** $-8xy$ and $16x$

   g $-28mn$ and $14n$ **h** $-18uv$ and $27vw$

4 Copy and complete each factorisation.

   a $-7a - 7b = -7(\underline{\quad})$ **b** $-3x - 12y = -3(\underline{\quad})$

   c $-24m - 8n = -8(\underline{\quad})$ **d** $-8b^2 - 4 = -4(\underline{\quad})$

   e $-6a + 3b = \underline{\quad}(2a - b)$ **f** $-49x - 7y = \underline{\quad}(7x + \underline{\ })$

   g $-9u + 18v = \underline{\quad}(u - \underline{\ })$ **h** $-6n^2 + 9 = \underline{\quad}(\underline{\ } - 3)$

5 Factorise each expression, then check your answer by expanding it.

   a $-3a - 18$ **b** $-12p - 24$ **c** $-5n - 20$

   d $-6y - 36$ **e** $-12a - 84$ **f** $-18x + 36$

   g $-21a + 63$ **h** $-6c - 24$ **i** $-25a + 75$

   j $-15x - 50y$ **k** $-a^2 - ab$ **l** $-xy + yb$

   m $-mn + mn^2$ **n** $-9p - 27pq$ **o** $-xyz + 8xy^2$

6 Copy and complete each factorisation.

   a $-2a^2 - 6ab = -2a(\underline{\quad})$ **b** $-4xy - 20y^2 = -4y(\underline{\quad})$

   c $-8mno + 16noq = -8no(\underline{\quad})$ **d** $-48b^2 - 16bc = \underline{\quad}(3b + \underline{\ })$

   e $-9ab + 27bc = \underline{\quad}(a - \underline{\ })$ **f** $-75x^2 - 25xy = \underline{\quad}(3x + \underline{\ })$

7 Factorise each expression, then check your answer by expanding.

   a $-3ap - 15pq$ **b** $-8rs - 20st$

   c $-7xy - 35y^2$ **d** $-15abc + 30bc$

   e $-8m^2 + 64mn$ **f** $-44b^2 - 11ab$

   g $-6fg + 15g^2$ **h** $-3xyz - 12uvx$

   i $-7a^2\,b - 28b^2c$ **j** $-8w^2 + 24uvw$

   k $-9r^2t - 18tv^2$ **l** $-12mn^2 + 48nm^2$

8 Factorise $-36a^2bc - 18b^2c + 72abc$, then check your answer by expanding it.

WORD SCRAMBLE

Unscramble each word to form words from this chapter.

DANPEX	GEBICLARA	NESSEPOIXR	KIEL
PLIMFSIY	BUSISTUTET	KUNLIE	MESRT
ECRSIFATO	IVENGEAT	GADIND	MERROPUNLA
LYPGINULMTI	EFFTENIOCCI	VIDGINDI	LUMOFRA
TACTGINBUSR			

Part A General topics

Calculators are not allowed.

1 Write Pythagoras' theorem for this triangle.

2 Write the formula for the area of a circle with radius r.

3 Convert 0.015 to a percentage.

4 How many fortnights are there in one year?

5 Evaluate $\sqrt[3]{27}$.

6 Joel works from 9 a.m. to 5 p.m. on Monday at $18.50 per hour. Calculate his pay.

7 Evaluate $(-8)^2$.

8 Find the value of d if the perimeter of this triangle is 58 m.

9 Expand $-2(3a + 9)$.

10 Find the mode of 6, 3, 2, 6, 5, 5, 6.

Part B Algebra

Calculators are allowed.

4-01 From words to algebraic expressions

11 Convert 'twice the sum of x and y' into an algebraic expression. Select the correct answer **A, B, C** or **D**.

  A $x + y$ **B** $2x + y$ **C** $2(x + y)$ **D** $x + 2y$

12 Find an algebraic expression for the area of a square of side $4a$. Select **A, B, C** or **D**.

  A $4a^2$ **B** $16a$ **C** $8a$ **D** $16a^2$

4-02 Substitution

13 If $a = -6$ and $b = 0.3$, evaluate each expression.

  a $2ab$ **b** $12 - 2a + b$ **c** $2a^2 - 5b$

4-03 Adding and subtracting terms

14 Simplify $3x - 5y - 2x + 8y$. Select **A, B, C** or **D**.

  A $x - 13y$ **B** $x + 3y$ **C** $5x + 3y$ **D** $-5x - 13y$

15 Simplify each expression.

  a $-3a + 4b - 8a$ **b** $16xy - 4x - y + 2yx$ **c** $8mn - m^2 - 12nm$

4-04 Multiplying and dividing terms

16 Simplify each expression.

    a $-4x \times (-3y)$ **b** $-5a^2 \times 6ab$ **c** $-9uv \times (-7vw)$

17 Simplify each expression.

    a $36ab \div (-4b)$ **b** $56x^2 \div 7x$ **c** $\dfrac{-15bc}{-25ac}$

18 Find a simplified algebraic expression for the area of this rectangle.

4-05 Expanding expressions

19 Is each equation true or false?

    a $5(3a - 1) = 15a - 1$ **b** $-4(3x + 2) = -12x - 8$ **c** $5a(a - 4) = 5a^2 - 4a$

20 Expand and simplify each expression.

    a $2(5m - 1) - (m + 3)$ **b** $a(2a + 1) - 3a(6a + 2)$

4-06 Factorising expressions

21 Factorise each expression.

    a $8a - 12b$ **b** $8u^2 - 28uv$ **c** $32ab^2 - 16a^2b$

4-07 Factorising with negative terms

22 Factorise each expression.

    a $-4m - 20n$ **b** $-30x + 18xy$ **c** $-25u^2 - 75uvw$

PYTHAGORAS' THEOREM

5

IN THIS CHAPTER YOU WILL:

- understand and write Pythagoras' theorem for right-angled triangles
- use Pythagoras' theorem to find the length of the hypotenuse or shorter side in a right-angled triangle, giving the answer as a surd or a decimal approximation
- use Pythagoras' theorem to test whether a triangle is right-angled
- investigate Pythagorean triads
- solve problems involving Pythagoras' theorem

Shutterstock.com/marekuliasz

WORDBANK

right-angled triangle A triangle with one angle that is exactly 90°. This angle is called the right angle.

hypotenuse The longest side of a right-angled triangle, the side opposite the right angle.

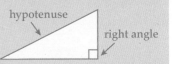

Pythagoras' theorem The rule or formula $c^2 = a^2 + b^2$ that relates the lengths of the sides of a right-angled triangle. Pythagoras was the ancient Greek mathematician who discovered this rule ('theorem' means rule).

PYTHAGORAS' THEOREM

In any right-angled triangle, the square of the hypotenuse is equal to the sum of the squares of the other two sides.

In the diagram, c is the length of the hypotenuse (longest side), and Pythagoras' theorem is $c^2 = a^2 + b^2$

EXAMPLE 1

State Pythagoras' theorem for each right-angled triangle.

a

b
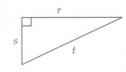

SOLUTION

Pythagoras' theorem is: Hypotenuse² = sum of squares of other two sides

a $q^2 = p^2 + r^2$

b $t^2 = r^2 + s^2$

✱ Remember, Pythagoras' theorem begins with (hypotenuse)², and q is the hypotenuse here.

EXAMPLE 2

Test Pythagoras' theorem on each triangle.

a
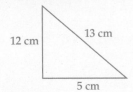

b
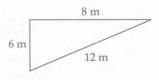

SOLUTION

a For Pythagoras' theorem to be true, hypotenuse² = sum of squares of the other two sides.
Does $13^2 = 5^2 + 12^2$?
    $169 = 25 + 144$ is true, so Pythagoras' theorem is true.
This means that the triangle is right-angled.

b Does $12^2 = 6^2 + 8^2$?
    $144 = 36 + 64$ is false, so Pythagoras' theorem is false here.
This means that the triangle is **not** right-angled.

Developmental Mathematics Book 4

ISBN 9780170351058

1   Which description is correct about the hypotenuse in a right-angled triangle? Select the correct answer **A, B, C** or **D**.

   A   the shortest side   B   next to the right angle

   C   the middle side   D   opposite the right angle

2   If the sides of a right-angled triangle have lengths x, y and z, with the hypotenuse being y, what is Pythagoras' theorem for the triangle? Select **A, B, C** or **D**.

   A $x^2 = y^2 + z^2$ B $y^2 = x^2 - z^2$

   C $y^2 = x^2 + z^2$ D $z^2 = x^2 + y^2$

3   Write Pythagoras' theorem for each right-angled triangle.

   a 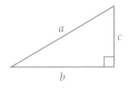   b 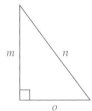   c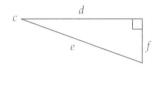

4   Write the length of the hypotenuse in each right-angled triangle.

   a 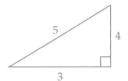   b 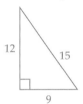   c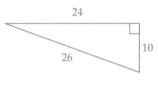

5   Write Pythagoras' theorem for each triangle in Question 4.

6   Test Pythagoras' theorem on each triangle.

   a 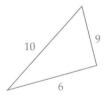   b   c

7   Which triangles in Question 6 are right-angled?

WORDBANK

surd A square root whose answer is not an exact number. For example, $\sqrt{8} = 2.8284...$ is a surd because there isn't an exact number squared that is equal to 8. As a decimal, the digits of $\sqrt{8}$ run endlessly without any repeating pattern.

exact form When the answer is written as an exact number, such as a whole number, decimal or a surd, and not rounded.

To find the length of the hypotenuse in a right-angled triangle:
■ write down Pythagoras' theorem in the form $c^2 = a^2 + b^2$ where c is the length of the hypotenuse
■ solve the equation
■ check that your answer is the longest side

EXAMPLE 3

Find the length of the hypotenuse in each triangle, writing your answer in exact form.

a
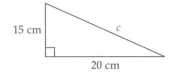

b
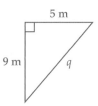

SOLUTION

a $\quad c^2 = a^2 + b^2$
$\quad\quad = 15^2 + 20^2$
$\quad\quad = 625$
$\quad c = \sqrt{625}$
$\quad c = 25$ cm ◄――――― This is in exact form.

b $\quad c^2 = a^2 + b^2$
$\quad q^2 = 5^2 + 9^2$
$\quad\quad = 106$
$\quad q = \sqrt{106}$ m ◄――――― This is in exact surd form.

 ✱ From the diagram, a hypotenuse of length 25 cm looks reasonable. It is also the longest side.

EXAMPLE 4

Find the value of d, correct to 2 decimal places.

SOLUTION

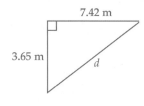

$c^2 = a^2 + b^2$
$d^2 = 3.65^2 + 7.42^2$
$\quad = 68.3789$
$d = \sqrt{68.3789}$
$\quad = 8.2691...$
$\quad \approx 8.27$ m

EXERCISE 5–02

1 What is Pythagoras' theorem for this triangle?
   Select the correct answer **A, B, C** or **D**.

   A $12^2 = 5^2 + c^2$ **B** $c^2 = 12^2 - 5^2$
   C $c^2 = 12^2 + 5^2$ **D** $5^2 = 12^2 + c^2$

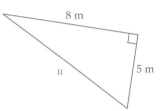

2 Find the value of c in Question **1**. Select **A, B, C** or **D**.

   A 13   **B** 10.9   **C** 119   **D** 11

3 Copy and complete this solution to find the value of u in
   this triangle correct to one decimal place.

   $u^2 = 5^2 + \underline{}^2$

   $u^2 = \underline{}$

   $u = \sqrt{\underline{}}$

   $\approx \underline{}$

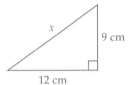

4 Find the length of the hypotenuse in each triangle below. Answer in exact form.

   a

   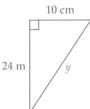

   b

   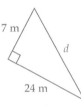

   c

   d

   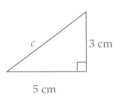

   e

   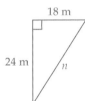

   f

   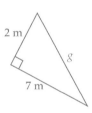

5 Round your answers to Question **4 d** and **f** to one decimal place.

6 Find the length of the hypotenuse in each triangle below. Answer correct to 2 decimal places.

   a

   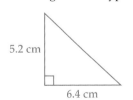

   b

   c

   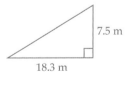

7 Find the length of the roof l correct to one decimal place.

   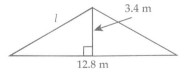

To find the length of a shorter side in a right-angled triangle:

- write down Pythagoras' theorem in the form $c^2 = a^2 + b^2$ where c is the length of the hypotenuse
- re-arrange the equation so that the shorter side is on the LHS (left-hand side)
- solve the equation
- check that your answer is shorter than the hypotenuse

EXAMPLE 5

Find the length of the unknown side in each triangle below. Answer in exact form.

a

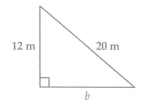

b

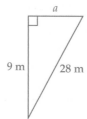

SOLUTION

a $20^2 = b^2 + 12^2$ ◄——— 20 is the hypotenuse
$b^2 + 12^2 = 20^2$ ◄——— Rearranging equation so that
$b^2 = 20^2 - 12^2$ b is on the LHS
$ = 256$
$b = \sqrt{256}$
$ = 16$ m

b $28^2 = a^2 + 9^2$
$a^2 + 9^2 = 28^2$
$a^2 = 28^2 - 9^2$ ◄——— Subtracting 9^2
$ = 703$
$a = \sqrt{703}$ m

＊   From the diagram, a length of 16 m looks reasonable. It is also shorter than the hypotenuse, 20 m.

EXAMPLE 6

Find, correct to one decimal place, the length of the unknown side in each triangle.

a

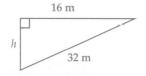

b

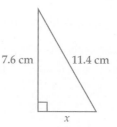

SOLUTION

a $32^2 = h^2 + 16^2$
$h^2 + 16^2 = 32^2$
$h^2 = 32^2 - 16^2$
$ = 768$
$h = \sqrt{768}$
$ = 27.7128...$
$ \approx 27.7$ m

b $11.4^2 = x^2 + 7.6^2$
$x^2 + 7.6^2 = 11.4^2$
$x^2 = 11.4^2 - 7.6^2$
$ = 72.2$
$x = \sqrt{72.2}$
$ = 8.4970...$
$ \approx 8.5$ cm

1 Find the value of b. Select the correct answer **A, B, C** or **D**.

   A 12   **B** 100

   C 35.3   **D** 10

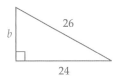

2 Find the value of each pronumeral in exact form.

   a

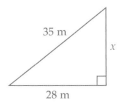

   b

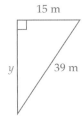

   c

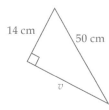

   d

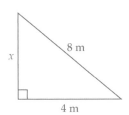

   e

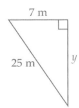

   f

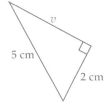

3 Find the value of each pronumeral, correct to 2 decimal places.

   a

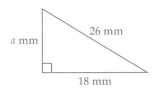

   b

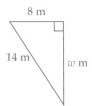

   c

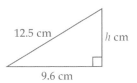

   d

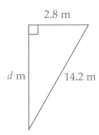

4 Find the height of the roof h as shown in the diagram below.

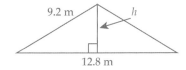

Mixed problems

To find the length of the hypotenuse, use $c^2 = a^2 + b^2$ where c is the hypotenuse.

To find the length of one of the shorter sides, use $b^2 = c^2 - a^2$ to find side b or $a^2 = c^2 - b^2$ to find side a.

EXAMPLE 7

Find the length of the unknown side in each triangle below. Leave your answer in exact form.

a

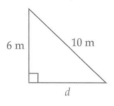

6 m   10 m

d

b

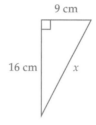

9 cm

16 cm x

c

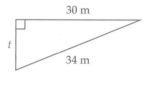

30 m

t

34 m

SOLUTION

a d is a shorter side
Square and subtract

$$d^2 = 10^2 - 6^2$$
$$= 100 - 36$$
$$= 64$$
$$d = \sqrt{64}$$
$$= 8 \text{ m}$$

b x is the hypotenuse
Square and add

$$x^2 = 9^2 + 16^2$$
$$= 81 + 256$$
$$= 337$$
$$x = \sqrt{337} \text{ cm}$$

c t is a shorter side
Square and subtract

$$t^2 = 34^2 - 30^2$$
$$= 1156 - 900$$
$$= 256$$
$$t = \sqrt{256}$$
$$= 16 \text{ m}$$

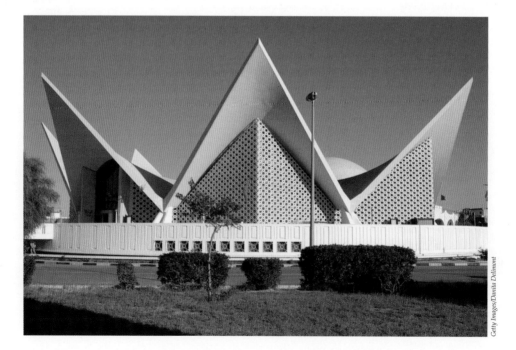

Getty Images/Danita Delimont

1   To find the length of the hypotenuse, x, in the triangle below, which rule is correct? Select
    the correct answer **A**, **B**, **C** or **D**.

    A $y^2 = x^2 + z^2$

    B $x^2 = y^2 - z^2$

    C $z^2 = y^2 - x^2$

    D $x^2 = y^2 + z^2$

2   To find the length of the shorter side, t, in the triangle
    below, which rule is correct?

    A $t^2 = r^2 + s^2$

    B $t^2 = r^2 - s^2$

    C $r^2 = t^2 - s^2$

    D $r^2 = s^2 + t^2$

3   Find the length of the unknown side in each triangle.
    Leave your answers in exact form.

    a
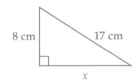

    b

    c
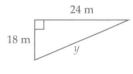

    d
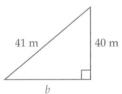

    e
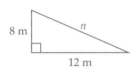

    f
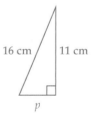

4   Find, correct to two decimal places, the value of each pronumeral.

    a
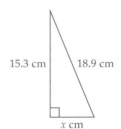

    b
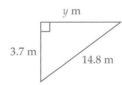

    c
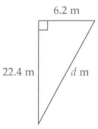

5   Jessica is making a triangular garden bed as shown.

    a   Find the value of b, correct to 1 decimal place.

    b   Calculate the perimeter of this garden.

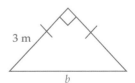

If the sides of a triangle follow the rule $c^2 = a^2 + b^2$, then the triangle must be right-angled. This is called the **converse** of Pythagoras' theorem, the theorem used in reverse.

> **To prove that a triangle is right-angled:**
> - substitute the lengths of its sides into the rule $c^2 = a^2 + b^2$
> - if it is true, then the triangle is right-angled
> - if it is false, then the triangle is not right-angled

EXAMPLE 8

Test whether each triangle is right-angled.

a

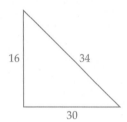

b

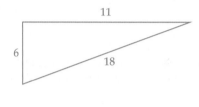

SOLUTION

Substitute the lengths of the sides into the rule $c^2 = a^2 + b^2$.

a Does $34^2 = 16^2 + 30^2$?

$1156 = 256 + 900$

$1156 = 1156$ ⟵——— Yes

This triangle is right-angled.

✱ The right angle is opposite the hypotenuse, 34, between the sides marked 16 and 30.

b Does $18^2 = 6^2 + 11^2$?

$324 = 36 + 121$

$324 \neq 157$ ⟵——— No

This triangle is **not** right-angled.

Image Source / Alamy

1 In the triangle below, which angle is the right angle? Select the correct angle **A**, **B** or **C**.

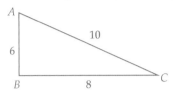

2 What is Pythagoras' theorem for the triangle above? Select **A**, **B** or **C**.

A $8^2 = 6^2 + 10^2$ **B** $6^2 = 8^2 + 10^2$ **C** $10^2 = 6^2 + 8^2$

3 Copy and complete to test if this triangle is right-angled:

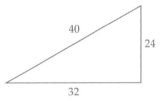

Does $40^2 = 24^2 + \underline{\quad}^2$?

$1600 = \underline{\quad} + 1024$

$1600 = \underline{\quad} \quad \underline{\quad}$

So the triangle _____ right-angled.

4 Test whether each triangle is right-angled.

a

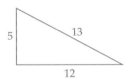

b

c

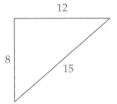

d

e

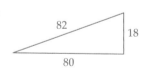

f

5 For each triangle that is right-angled in Question **4**, name the two sides the right angle is between.

6 A ladder 6.1 m long is leant against a wall 1.1 m from its base.

a Draw a diagram to show this information.

b Will the ladder reach a window 6 m high?

WORDBANK

Pythagorean triad A set of 3 numbers that follows Pythagoras' theorem, $c^2 = a^2 + b^2$.

To prove a group of 3 numbers is a Pythagorean triad:
- ■ substitute the numbers into the rule $c^2 = a^2 + b^2$, where c is the largest number
- ■ if it is true, then the numbers form a Pythagorean triad
- ■ if it is false, then the numbers do not form a Pythagorean triad

EXAMPLE 9

Test whether each set of numbers form a Pythagorean triad.

a {5, 12, 13} **b** {9, 14, 15}

SOLUTION

a Does $13^2 = 5^2 + 12^2$?

✱ Always substitute the largest number for c.

$169 = 25 + 144$
$169 = 169$ true

So {5, 12, 13} is a Pythagorean triad.

b Does $15^2 = 9^2 + 14^2$?
$225 = 81 + 196$
$225 = 277$ false

So {9, 14, 15} is not a Pythagorean triad.

If the three numbers in a Pythagorean triad are multiplied by the same value, then the new numbers also form a Pythagorean triad.

EXAMPLE 10

If {5, 12, 13} is a Pythagorean triad, find two more Pythagorean triads using multiples of {5, 12, 13}.

SOLUTION

$\{5 \times 2, 12 \times 2, 13 \times 2\} = \{10, 24, 26\}$ ←——————— Multiplying {5, 12, 13} by 2.

✱ Check that $26^2 = 10^2 + 24^2$

$\{5 \times 3, 12 \times 3, 13 \times 3\} = \{15, 36, 39\}$ ←——————— Multiplying {5, 12, 13} by 3.

1   Which set of numbers is a Pythagorean triad? Select the correct answer **A, B, C** or **D**.

    A {3, 4, 4}   **B** {3, 4, 5}

    C {3, 4, 6}   **D** {3, 4, 7}

2   To prove that a set of numbers forms a Pythagorean triad using the formula $c^2 = a^2 + b^2$, which phrase describes c? Select **A, B, C** or **D**.

    A the smallest number   **B** the largest number

    C the middle number   **D** the square number

3   Which set of numbers is a Pythagorean triad? Select **A, B, C** or **D**.

    A {6, 8, 9}   **B** {8, 6, 5}

    C {6, 9, 10}   **D** {6, 8, 10}

4   Copy and complete:

    $25^2 = 7^2 + 24^2$
    $625 = \_\_\_ + 576$
    $625 = \_\_\_\_$

    So {7, 24, 25} _____ a Pythagorean triad.

5   Test whether each set of numbers form a Pythagorean triad.

    a {5, 12, 13}   **b** {9, 12, 16}

    c {9, 15, 17}   **d** {9, 12, 15}

    e {11, 30, 31}   **f** {9, 20, 25}

    g {15, 20, 25}   **h** {8, 15, 17}

    i {10, 24, 26}   **j** {7, 24, 25}

    k {9, 40, 41}   **l** {9, 14, 16}

6   For each Pythagorean triad given, find two more Pythagorean triads by multiplying by the same number.

    a {3, 4, 5}   **b** {8, 15, 17}   **c** {9, 40, 41}   **d** {5, 12, 13}

7   **a**   Prove that {14, 48, 50} is a Pythagorean triad.

    b   Which Pythagorean triad is this a multiple of?

8   Is each statement true or false?

    a   {60, 80, 100} is a Pythagorean triad.

    b   {12, 16, 25} is a multiple of {3, 4, 5}.

9   Draw a right-angled triangle with sides 7 cm, 24 cm and 25 cm. Is {7, 24, 25} a Pythagorean triad?

To solve a problem using Pythagoras' theorem:
- draw a diagram if it is not given and draw a right-angled triangle
- identify the unknown value
- use $c^2 = a^2 + b^2$ to solve an equation
- answer the problem in words.

EXAMPLE 11

Lachlan drove 5 km east and then 7 km south to stop by Dean's house. When returning home, he went the quickest way, in a straight line. How far, correct to one decimal place, was the direct way?

5 km east

Quickest way home (d km)

7 km south

Dean's house

SOLUTION

Let d km be the direct way home.

$d^2 = 5^2 + 7^2$ ⟵——— d is the hypotenuse of the triangle
$\quad = 74$
$d = \sqrt{74}$
$\quad = 8.6023…$
$\quad \approx 8.6$

So the shortest distance home for Lachlan is 8.6 km, rounded to 1 decimal place.

EXAMPLE 12

Taylor uses a 2.5 m support wire to hold up her rose bush. How tall (correct to one decimal place) is the bush if the wire is 1.8 m from the base of the plant and the foliage is 60 cm high?

SOLUTION

Let the height of the rose bush (not including the flowers) be h m.

$h^2 = 2.5^2 - 1.8^2$ ⟵——— h is a shorter side
$\quad = 3.01$
$h = \sqrt{3.01}$
$\quad = 1.7349…$
$\quad \approx 1.7$

The height of the rose bush is 1.7 m + 60 cm = 1.7 m + 0.6 m = 2.3 m

60 cm

2.5 m

1.8 m

1   Raoul leans a ladder against a 8 m high wall
    so that it reaches the top of it. He places the
    ladder 1.9 m from the base of the wall. Which
    is the correct diagram for this situation?
    Select **A**, **B** or **C**.

    A

    8 m
    1.9 m

    B

    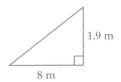

    1.9 m
    8 m

    C

    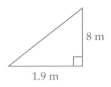

    8 m
    1.9 m

Shutterstock.com/Paul Wishart

2   Find, correct to two decimal places, the length of the ladder in Question **1**.

3   Renee leans a 7.4 m ladder against a wall so that it reaches the top of it. The end of the
    ladder is 2 m from the base of the wall. Find, correct to two decimal places, the height of
    the wall.

4   Find the length of the diagonal in the rectangle and square correct to the nearest whole
    number.

    a   4 cm   **b**

    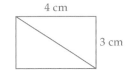

    3 cm

    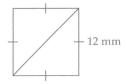

    12 mm

5   If *ABC* is an isosceles triangle, find its perimeter.

    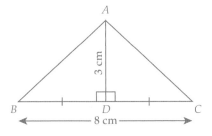

    A

    3 cm

    B   *D*   *C*
    ←——— 8 cm ———→

6   What length of timber is needed for the diagonal crosspiece of this gate?

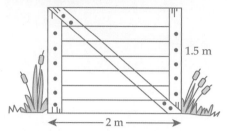

7   The wire supporting a tree is 13 m long and the bottom of the wire is 5 m from the base of the tree. How high up the tree does the wire reach?

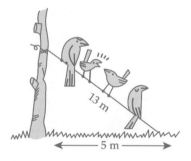

8   Will travelled 30 km in a northeast direction before travelling 12 km west so that he was due north of his starting point. How far north did he travel (to the nearest kilometre)?

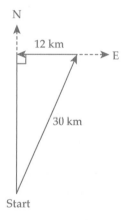

ISBN 9780170351058

PYTHAGORAS' THEOREM CROSSWORD

Make a copy of this puzzle then complete this crossword using the given clues.

Across

1. A mathematical word meaning rule or formula.
3. The ancient Greek mathematician who discovered the rule about right-angled triangles.
6. A shape with three straight sides.
7 and 9.   Another name for a 90° angle (two words)
8. Pythagoras' theorem describes the relationship between the three _____ of a right-angled triangle.
10. The opposite of squaring a number (two words)

Down

2. The longest side of a right-angled triangle.
4 and 5.   {3, 4, 5} is one example of a set of three numbers called this (two words).

Part A General topics

Calculators are not allowed.

1 Write an algebraic expression for the number that is 6 more than r.

2 Convert 5.8% to a decimal.

3 Expand $9y(y - 2)$.

4 How many degrees in a revolution?

5 Evaluate $13.26 \div 6$.

6 Find the median of 6, 3, 2, 6, 5, 4, 6.

7 Does the point $(-3, 0)$ lie on the x-axis or y-axis?

8 Find the volume of this prism.

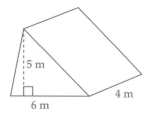

9 Evaluate 39×11.

10 Write a simplified algebraic expression for the perimeter of this rectangle.

Part B Pythagoras' theorem

Calculators are allowed.

5–01 Pythagoras' theorem

11 What is Pythagoras' theorem for this triangle? Select the correct answer **A, B, C** or **D**.

A $z^2 = x^2 + y^2$ **B** $y^2 = x^2 + z^2$
C $z^2 = y^2 - x^2$ **D** $x^2 = y^2 + z^2$

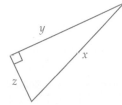

5–02 Finding the hypotenuse

12 Find the length of the hypotenuse. Select **A, B, C** or **D**.

A 17 cm   **B** 13 cm
C 11 cm   **D** 15 cm

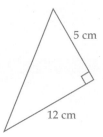

Developmental Mathematics Book 4

ISBN 9780170351058

13 Find the value of each pronumeral, giving your answer in exact form.

a

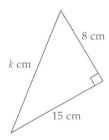

b

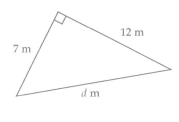

5-03 Finding a shorter side

14 Find, correct to two decimal places, the value of each pronumeral.

a

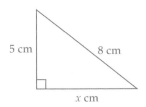

b

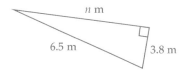

5-04 Mixed problems

15 Find, correct to one decimal place, the value of each pronumeral.

a

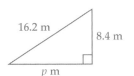

b

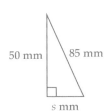

c

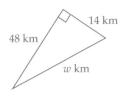

5-05 Testing for right-angled triangles

16 Test whether each triangle is right-angled.

a

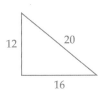

b

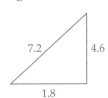

5-06 Pythagorean triads

17 Test whether each set of numbers is a Pythagorean triad.

    a {18, 80, 82}   **b** {10, 24, 28}

5–07 Pythagoras' theorem problems

18 Find, correct to one decimal place, the length of the longest umbrella that can fit inside a suitcase measuring 1.5 m long and 0.6 m wide.

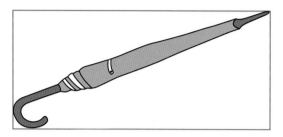

19 Find the perimeter of this trapezium.

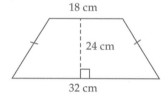

TRIGONOMETRY

6

WHAT'S IN CHAPTER 6?

IN THIS CHAPTER YOU WILL:

- label the sides of a right-angled triangle: opposite, adjacent, hypotenuse
- learn the trigonometric ratios for right-angled triangles: sine (sin), cosine (cos), tangent (tan)
- use trigonometric ratios to find unknown sides in right-angled triangles
- use trigonometric ratios to find unknown angles in right-angled triangles
- use bearings to describe direction and solve problems involving bearings

Shutterstock.com/Mimadeo

WORDBANK

trigonometry The study of the measurement of sides and angles in triangles.

hypotenuse The longest side of a right-angled triangle, the side opposite the right angle.

opposite side The side facing a given angle in a right-angled triangle.

adjacent side The side next to a given angle in a triangle leading to the right angle.

For angle C in this right-angled triangle, AB is the opposite side, BC is the adjacent side and AC is the hypotenuse.

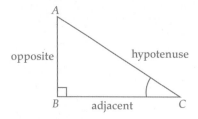

EXAMPLE 1

Name the hypotenuse, opposite and adjacent sides for the marked angle in each right-angled triangle.

a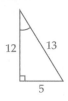

b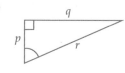

SOLUTION

a Hypotenuse = 13 ⟵——— opposite the right angle

   Opposite side = 5 ⟵ ——— opposite the marked angle

   Adjacent side = 12 ⟵——— next to the marked angle

b Hypotenuse = r

   Opposite side = q

   Adjacent side = p

When **labelling triangles:**
- use capital letters for angles: P, Q, R
- use small letters for sides: p, q, r
- use the same letter for sides and angles opposite each other

side p is opposite angle P

side q is opposite angle Q

side r is opposite angle R

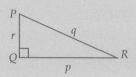

1 In a right-angled triangle, what is the hypotenuse? Select the correct answer **A, B, C** or **D**.

  A the shortest side
  B the side opposite the marked angle
  C the side opposite the right angle
  D the side adjacent to the marked angle

2 Which phrase describes the side a in $\triangle ABC$? Select **A, B, C** or **D**.

  A opposite the right angle
  B opposite angle A
  C adjacent to angle A
  D opposite the hypotenuse

3 Name the hypotenuse, opposite and adjacent sides for the marked angle in this right-angled triangle.

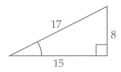

4 Copy each triangle and complete the labelling of each side and angle.

  a 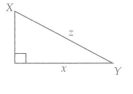　**b** 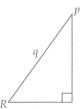　**c**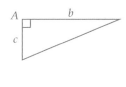

5 For each triangle in Question **4**, name the hypotenuse.

6 For each triangle, name the hypotenuse, opposite and adjacent sides in relation to the marked angle.

  a 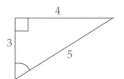　**b** 　**c**

  d 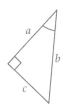　**e** 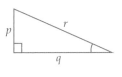　**f**

7 Draw a right-angled triangle and label all sides and angles using your choice of letters.

WORDBANK

trigonometric ratio A ratio such as sine, cosine and tangent that compares two sides of a right-angled triangle.

theta, θ A Greek letter often used to represent angles.

There are three basic trigonometric ratios for right-angled triangles: **sine**, **cosine** and **tangent**, which are abbreviated **sin**, **cos** and **tan** respectively.

$$\sin\theta = \frac{\text{opposite}}{\text{hypotenuse}} \qquad \cos\theta = \frac{\text{adjacent}}{\text{hypotenuse}} \qquad \tan\theta = \frac{\text{opposite}}{\text{adjacent}}$$

This phrase may help you to remember the formulas.
Super **O**ld **H**eroes **C**an't **A**lways **H**ide **T**heir **O**wn **A**ge.
The initials in bold, **SOH CAH TOA**, give the initials of the sin, cos and tan ratios.

EXAMPLE 2

Write the values of sin A, cos A and tan A for this triangle.

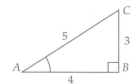

SOLUTION

For angle A, opposite = 3, adjacent = 4 and hypotenuse = 5, so:

$$\sin A = \frac{\text{opposite}}{\text{hypotenuse}} \qquad \cos A = \frac{\text{adjacent}}{\text{hypotenuse}} \qquad \tan A = \frac{\text{opposite}}{\text{adjacent}}$$

$$= \frac{3}{5} \qquad\qquad = \frac{4}{5} \qquad\qquad = \frac{3}{4}$$

EXAMPLE 3

For $\triangle XYZ$, write the expression for sin X, cos X and tan X.

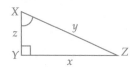

SOLUTION

For $\angle X$, opposite = x, adjacent = z and hypotenuse = y, so:

$$\sin X = \frac{\text{opposite}}{\text{hypotenuse}} \qquad \cos X = \frac{\text{adjacent}}{\text{hypotenuse}} \qquad \tan X = \frac{\text{opposite}}{\text{adjacent}}$$

$$= \frac{x}{y} \qquad\qquad = \frac{z}{y} \qquad\qquad = \frac{x}{z}$$

1 For $\triangle ABC$, find the value of cos B. Select the correct answer
   A, **B**, **C** or **D**.

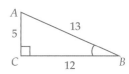

   A $\dfrac{12}{13}$ **B** $\dfrac{5}{12}$ **C** $\dfrac{5}{13}$ **D** $\dfrac{12}{5}$

2 For $\triangle ABC$ in Question **1**, find the value of cos A.
   Select **A**, **B**, **C** or **D**.

   A $\dfrac{12}{13}$ **B** $\dfrac{5}{12}$ **C** $\dfrac{5}{13}$ **D** $\dfrac{12}{5}$

3 Copy and complete each formula.

   a $\sin\theta = \dfrac{\rule{2cm}{0.4pt}}{\text{hypotenuse}}$ **b** $\cos\theta = \dfrac{\text{adjacent}}{\rule{2cm}{0.4pt}}$ **c** $\tan\theta = \dfrac{\rule{2cm}{0.4pt}}{\text{adjacent}}$

4 For this triangle, find as a simplified fraction:

   a $\sin P$, $\cos P$, $\tan P$

   b $\sin Q$, $\cos Q$ and $\tan Q$

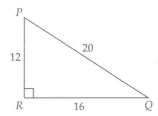

6 For each marked angle, find:

   a the sin ratio   **b** the cos ratio   **c** the tan ratio

   i

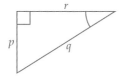

   ii

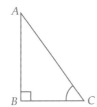

7 Copy and complete each statement for $\triangle DEF$.

   a $\sin F = \dfrac{7}{\square}$ **b** $\cos F = \dfrac{\square}{25}$

   c $\tan F = \dfrac{\square}{24}$ **d** $\sin D = \dfrac{\square}{25}$

   e $\cos D = \dfrac{7}{\square}$ **f** $\tan D = \dfrac{\square}{7}$

8 Draw a right-angled triangle in which $\sin A = \dfrac{40}{41}$, $\cos A = \dfrac{9}{41}$ and $\tan A = \dfrac{40}{9}$.

The tangent ratio can be used to find unknown sides in right-angled triangles.

$$\tan\theta = \frac{\text{opposite}}{\text{adjacent}}$$

EXAMPLE 4

Find the value of b correct to 2 decimal places.

SOLUTION

$$\tan 35° = \frac{\text{opposite}}{\text{adjacent}}$$

$$\tan 35° = \frac{b}{12}$$ ←——— b is opposite to 35° and 12 is adjacent.

$$12 \tan 35° = b$$ ←——— Multiplying both sides by 12

$$b = 8.4024...$$ ←——— On a calculator, enter: 12 35 ▬

Make sure that your calculator is in degrees (D) mode

$$b \approx 8.40$$

✳ From the diagram, $b \approx 8.40$ looks reasonable.

EXAMPLE 5

Find the value of x correct to 1 decimal place.

SOLUTION

$$\tan 42° = \frac{\text{opposite}}{\text{adjacent}}$$

$$\tan 42° = \frac{16}{x}$$ ←——— 16 is opposite 42° and x is adjacent.

$$x \tan 42° = 16$$ ←——— Multiplying both sides by x.

$$x = \frac{16}{\tan 42°}$$ ←——— Dividing both sides by tan 42°.

$$x = 17.7698...$$ ←——— On a calculator, enter: 16 42 ▬

$$x \approx 17.8$$

✳ From the diagram, $x \approx 17.8$ looks reasonable.

Note that we can jump from the 2nd to 4th lines of working by swapping the places of

tan 42° and x:

$$\tan 42° = \frac{16}{x} \text{ becomes } x = \frac{16}{\tan 42°}.$$

EXERCISE 6–03

1   Write the correct expression for tan 28° for this triangle.
    Select the correct answer **A**, **B**, **C** or **D**.

    A $\tan 28° = \dfrac{a}{9}$ **B** $\tan 28° = \dfrac{a}{14}$

    C $\tan 28° = \dfrac{9}{a}$ **D** $\tan 28° = \dfrac{14}{a}$

    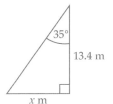

2   Evaluate 6.4 tan 36° correct to 2 decimal places. Select **A**, **B**, **C** or **D**.

    A 4.64   **B** 46.49   **C** 12.56   **D** 4.65

3   Copy and complete this solution to find *m*, correct to two decimal places.

    $\tan 73° = \dfrac{m}{8}$

    ___ tan 73° = *m*

    m = _____

    m ≈ _____

4   Find, correct to 2 decimal places, the value of each pronumeral.

    a   **b**    **c**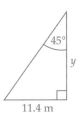

5   Find, correct to 1 decimal place, the value of each pronumeral.

    a   **b** 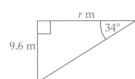   **c**

6   Helena is standing 18 m away from the base of an
    apartment building. She looks up to the top of
    the building through an angle of 24°. How high is
    the building, to 2 decimal places?

    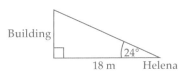

7   A tower is 45 m high. Josh is standing level with the base of
    the tower and looks through an angle of 19° to see the top of
    the tower. How far is Josh standing from the base of the tower,
    to 1 decimal place?

    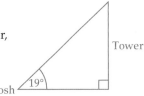

The tangent ratio can be used to find unknown angles in right-angled triangles.

$$\tan \theta = \frac{\text{opposite}}{\text{adjacent}}$$

EXAMPLE 6

Find the size of angle A in the triangle, correct to the nearest degree.

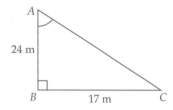

SOLUTION

$$\tan A = \frac{\text{opposite}}{\text{adjacent}}$$

$\tan A = \dfrac{17}{24}$ ⟵ 17 is opposite to angle A and 24 is adjacent.

$A = 35.3112\ldots$ ⟵ On a calculator, enter: SHIFT tan 17 a^b/$_c$ 24 =

$A \approx 35°$

✱ From the diagram, $A = 35°$ looks reasonable.

EXAMPLE 7

Construct $\triangle PQR$ where side p is 9 cm, side q is 12 cm and $\angle R$ is 90°. Calculate the size of $\angle Q$ correct to the nearest degree, and check your answer by measuring it with your protractor.

SOLUTION

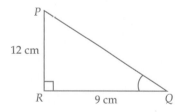

$\tan Q = \dfrac{12}{9}$ ⟵ 12 is opposite to $\angle Q$ and 9 is adjacent.

$Q = 53.1301\ldots$ ⟵ On a calculator, enter: SHIFT tan 12 a^b/$_c$ 9 =

$Q \approx 53°$

Measure $\angle Q$ with a protractor and check that it is 53°.

1   Find A if $\tan A = \dfrac{3}{8}$. Select the correct answer **A**, **B**, **C** or **D**.

    A 21°   **B** 6.5   **C** 20°   **D** 69°

2   Find $\tan Q$ for this triangle. Select **A**, **B**, **C** or **D**.

    A $\dfrac{7}{3}$ **B** $\dfrac{6}{3}$

    C $\dfrac{1}{2}$ **D** $\dfrac{3}{7}$

3   Is each solution true or false?

    a $\tan A = \dfrac{3}{5}$ **b** $\tan P = \dfrac{7}{3}$ **c** $\tan Y = \dfrac{2}{9}$ **d** $\tan W = \dfrac{17}{13}$

        $A \approx 31°$ $P \approx 68°$ $Y \approx 9°$ $W \approx 53°$

4   Find the size of $\angle A$ in each triangle, correct to the nearest degree.

    a   **b**   **c**

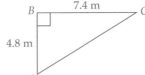

5   Construct ΔPQR with side $p = 8$ cm, side $q = 15$ cm and $\angle R = 90°$. Calculate the size of $\angle P$ correct to the nearest degree, and check your answer by measuring it with your protractor.

6   Find the value of θ in each triangle, correct to the nearest degree.

    a   **b**   **c**

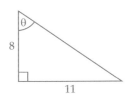
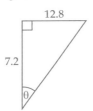
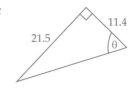

7   Brittany was rowing out to sea from the base of a cliff 32.5 m high. After rowing out 11.8 m, what angle must she look up to see the top of the cliff, to the nearest degree?

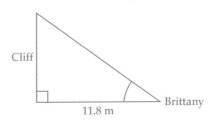

8   Mark was on the roof of a 42 m high tower. Monique was on a seat 16.9 m from the base of the tower. Mark called out to her. Through what angle will she need to look up to see Mark, to the nearest degree? (Draw a diagram first.)

To find an unknown side, select the correct trigonometric ratio depending on which side and angle is given in the problem.

$$\sin \theta = \frac{\text{opposite}}{\text{hypotenuse}}$$

$$\cos \theta = \frac{\text{adjacent}}{\text{hypotenuse}}$$

$$\tan \theta = \frac{\text{opposite}}{\text{adjacent}}$$

EXAMPLE 8

a Find b correct to 1 decimal place.

b Find r correct to 2 decimal places.

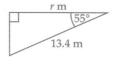

SOLUTION

a b is **opposite** 64° and 18 is the **hypotenuse**, so use **sin**.

$$\sin 64° = \frac{\text{opposite}}{\text{hypotenuse}}$$

$$\sin 64° = \frac{b}{18}$$

$$18 \sin 64° = b$$

$b = 16.1782 \dots$ ⟵——— On a calculator, enter: 18 ✕ sin 64 =

$b \approx 16.2$

✱ From the diagram, $b \approx 16.2$ looks reasonable.

b r is **adjacent** to 55° and 13.4 is the **hypotenuse**, so use **cos**.

$$\cos 55° = \frac{\text{adjacent}}{\text{hypotenuse}}$$

$$\cos 55° = \frac{r}{13.4}$$ ⟵——— r is adjacent to 55° and 13.4 is the hypotenuse.

$$13.4 \cos 55° = r$$

$r = 7.6859\dots$ ⟵——— On a calculator, enter: 13.4 ✕ cos 55 =

$r \approx 7.69$

EXERCISE 6-05

1 Which trigonometric ratio could be used to find side b?
Select the correct answer **A**, **B**, **C** or **D**.

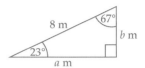

A sin 67° **B** sin 23°

C cos 23° **D** tan 67°

2 Which trigonometric ratio could be used to find side a in Question **1**? Select **A**, **B**, **C** or **D**.

A sin 67° **B** sin 23° **C** cos 67° **D** tan 23°

3 Copy and complete to find x correct to one decimal place.

$$\cos 42° = \frac{\quad}{19.4}$$

$$19.4 \underline{\quad\quad} = x$$

$$x = \underline{\quad\quad\quad\quad\quad}$$

$$x \approx \underline{\quad\quad}$$

4 Find the value of each pronumeral, correct to two decimal places.

a

b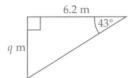

c

d

e

5 Hannah was looking up at a cat on the third floor of a building through an angle of 34° to the horizontal. She could see a distance of only 6 m in a straight line. How high (correct to two decimal places) above her was the cat?

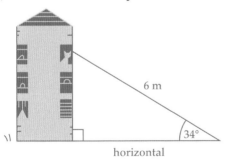

horizontal

6 Lyn is in a city square looking up through 38° to the top of an apartment block through a line of sight 45 m long. How high is the block, correct to one decimal place?

7 Ryan visits the Leaning Tower of Pisa, which is leaning at 7° to the vertical. The tower is 65 m tall, but the top of the tower is less than 65 m from the ground because of the lean. How high (correct to one decimal place) is the top of the tower from the ground?

EXAMPLE 9

a Find h correct to 1 decimal place.

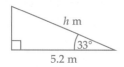

b Find q correct to 2 decimal places.

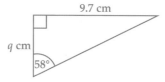

SOLUTION

a 5.2 is **adjacent** to 33° and h is the **hypotenuse**, so use **cos**.

$$\cos 33° = \frac{\text{adjacent}}{\text{hypotenuse}}$$

$$\cos 33° = \frac{5.2}{h}$$

$$h = \frac{5.2}{\cos 33°} \qquad \longleftarrow \text{ } h \text{ and } \cos 33° \text{ swap places.}$$

$$h = 6.2002\ldots \qquad \longleftarrow \text{ On a calculator, enter: 5.2 } \boxed{÷} \text{ } \boxed{\cos} \text{ 33 } \boxed{=}$$

$$h \approx 6.2$$

b 9.7 is **opposite** 58° and q is **adjacent** to 58°, so use **tan**.

$$\tan 58° = \frac{\text{opposite}}{\text{adjacent}}$$

$$\tan 58° = \frac{9.7}{q}$$

$$q = \frac{9.7}{\tan 58°} \qquad \longleftarrow \text{ } q \text{ and } \tan 58° \text{ swap places.}$$

$$q = 6.0612\ldots \qquad \longleftarrow \text{ On a calculator, enter: 9.7 } \boxed{÷} \text{ } \boxed{\tan} \text{ 58 } \boxed{=}$$

$$q \approx 6.06$$

Jeremy Inglis / Alamy

1   Which trigonometric ratio could you use to find side c? Select the correct answer **A, B, C** or **D**.

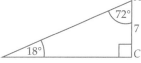

    A sin 18°   **B** sin 72°

    C cos 18°   **D** tan 72°

2   Which trigonometric ratio could you use to find side a in Question 1? Select **A, B, C** or **D**.

    A sin 18°   **B** cos 72°   **C** cos 18°   **D** tan 72°

3   If $\cos 53° = \dfrac{6}{c}$, write a formula for c.

4   Find, correct to 2 decimal places, the value of each pronumeral.

    a 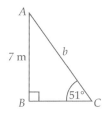   **b** 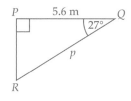   **c**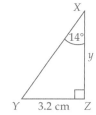

5   Copy and complete to find y, correct to one decimal place.

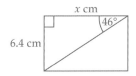

    $\tan 42° = \dfrac{2.8}{y}$

    $y = \dfrac{2.8}{}$

    $y = \underline{\hspace{3cm}}$

    $y \approx \underline{\hspace{2cm}}$

6   Find x, correct to two decimal places.

    a 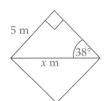   **b**

7   Mohammed's house has a roof that is inclined at 35° to the horizontal. The height of the roof at the top is 1.8 m. How far across is the top from the edge of the roof, to 2 decimal places?

To find an unknown angle, select the correct trigonometric ratio, depending on which sides are given in the problem.

$$\sin\theta = \frac{\text{opposite}}{\text{hypotenuse}} \qquad \cos\theta = \frac{\text{adjacent}}{\text{hypotenuse}} \qquad \tan\theta = \frac{\text{opposite}}{\text{adjacent}}$$

EXAMPLE 10

Find the size of each marked angle, correct to the nearest degree.

a

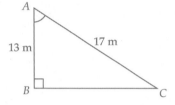

b

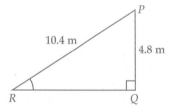

SOLUTION

a 13 is **adjacent** to $\angle A$ and 17 is the **hypotenuse**, so use **cos**.

$$\cos A = \frac{\text{adjacent}}{\text{hypotenuse}}$$

$$\cos A = \frac{13}{17}$$

$A = 40.1191\ldots$ ⟵——— On a calculator, enter: 13 17 =

$A \approx 40°$

b 4.8 is **opposite** $\angle R$ and 10.4 is the **hypotenuse**, so use **sin**.

$$\sin R = \frac{\text{opposite}}{\text{hypotenuse}}$$

$$\sin R = \frac{4.8}{10.4}$$

$R = 27.4864\ldots$ ⟵——— On a calculator, enter: 4.8 10.4 =

$R \approx 27°$

1 Which trigonometric ratio could be used to find $\angle A$? Select the correct answer **A**, **B** or **C**.

   A $\cos A$ **B** $\tan A$ **C** $\sin A$

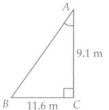

2 Which trigonometric ratio could be used to find $\angle T$? Select **A**, **B** or **C**.

   A $\cos T$ **B** $\tan T$ **C** $\sin T$

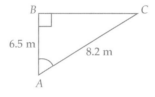

3 Copy and complete to find the size of $\angle R$, correct to the nearest degree.

   $\sin R = \dfrac{5}{7}$

   $R = \underline{\hspace{3cm}}$

   $R \approx \underline{\hspace{1.5cm}}$

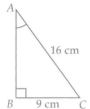

4 Find $\angle A$ in each triangle, correct to the nearest degree.

   a

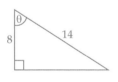

   b

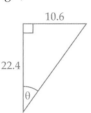

   c

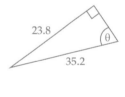

5 Find the value of θ in each triangle, correct to the nearest degree.

   a

   b

   c

6 Construct $\triangle PQR$ where side $p = 9$ cm, side $r = 14$ cm and $\angle R = 90°$. Calculate the size of $\angle Q$ correct to the nearest degree, then check that it is correct by measuring $\angle Q$ with your protractor.

7 Lauren placed a 6.8 m long ladder up to a window 4.2 m high. At what angle to the ground should she place the ladder in order for it to reach the window, to the nearest degree?

8 A rectangle has a length of 15.6 cm and its diagonal is 18.4 cm long. Find θ and α, to the nearest degree.

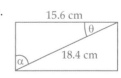

An angle of elevation is the angle looking **up** at an object, measured between the horizontal and the line of sight.

An angle of depression is the angle looking **down** at an object, measured between the horizontal and the line of sight.

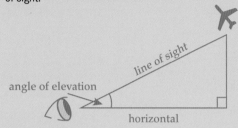

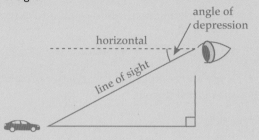

EXAMPLE 11

Amber saw the rooftop of a shopping complex through an angle of elevation of 27°. If she was 23 m from the base of the complex, how high was the complex?

SOLUTION

Need to find the height of the complex, h.

h is **opposite** 27° and 23 is **adjacent** to 27°, so use **tan**.

$$\tan 27° = \frac{h}{23}$$
$$23 \tan 27° = h$$

$h = 11.7190\ldots$

$h \approx 11.72$

The height of the complex is 11.72 m.

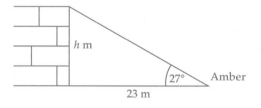

EXAMPLE 12

From the top of a 9 m hill, Dylan looks down to a lake at an angle of depression of 28°.

How far is the lake from the foot of the hill?

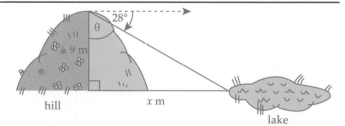

SOLUTION

The angle of depression is outside the triangle, but we can find θ because:

$\theta + 28° = 90°$

$\theta = 90° - 28°$

$\theta = 62°$

x is **opposite** θ and 9 is **adjacent** to 27°, so use **tan**.

$$\tan 62° = \frac{x}{9}$$
$$9 \tan 62° = x$$

$x = 16.9265\ldots$

$x \approx 16.9$

The lake is 16.9 m from the foot of the hill.

1   Which trigonometric ratio could be used to find angle B?
    Select the correct answer **A**, **B** or **C**.

    A $\cos B$ **B** $\tan B$ **C** $\sin B$

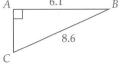

2   Is each marked angle an angle of elevation or angle of depression?

    a   **b**   **c**

3   Jordan flies a kite at an angle of elevation of 35°. How high is the kite above the ground, to 2 decimal places, if Jordan's arm is 1.5 m above the ground and the piece of string is 7.5 m long?

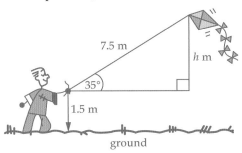

4   From a hill 12 m high, Hayley sees a lake through an angle of depression of 52°. How far away is the lake from the foot of the hill, to 2 decimal places?

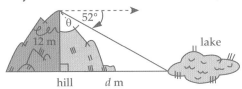

5   Abdul was on the fourth floor of his office and saw a garden through an angle of depression of 36° at 16 m from the base of the building. How high was Abdul above the ground, to 1 decimal place?

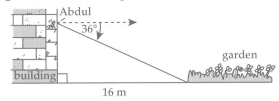

6   Jonah looks up at a hill 22 m high from a point 10 m from the base of the hill. Find the angle of elevation of the hill, to the nearest degree.

7   Suri placed a 4 m ladder at an angle of elevation of 62° to clean a window on the second floor of her house. If the window is 3.6 m above the ground, will the ladder reach?

8   Natalie sees a boat from the top of a cliff 32 m above sea level. The direct distance to the boat along her line of sight is 48 m. Find the angle of depression of the boat, to the nearest degree.

WORDBANK

bearing The direction of a place from another place, usually given as an angle.

compass rose A compass diagram showing north, south, east and west.

compass bearings The directions on the compass: the 16 main points are shown below.

three-figure bearings (or true bearings) Bearings using angles from 000° to 360° measured clockwise from north (000°) and using three digits, such as 125°.

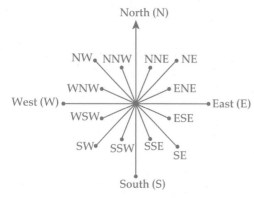

Compass bearings

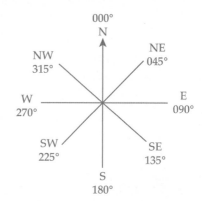

Three-figure bearing

- NE = north-east, SE = south-east, NW = north-west, SW = south-west
- NNW = north-northwest, WNE = west-northwest, WSW = west-southwest, SSW = south-southwest

---

EXAMPLE 13

Show each bearing on a compass rose.

a NE **b** WSW **c** 142° **d** 305°

SOLUTION

a

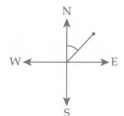

NE is halfway between north and east.

b

N, W, E, S diagram

WSW is halfway between west and southwest.

c

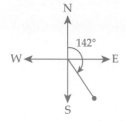

142° is between east (90°) and south (180°).

d

N, W, E, S diagram with 305°

305° is between west (270°) and north (360°).

EXAMPLE 14

Write each compass bearing as a three-figure bearing.

a NE   **b** WSW

SOLUTION

a NE is halfway between N and E, so NE is 45° from north.

NE is 045°

b WSW is halfway between W and SW, WSW is $\frac{1}{2} \times 45° = 22.5°$ from west.

West is 270°, so WSW = 270° − 22.5° = 247.5°.

EXERCISE 6-09

1 What is the three-figure bearing for the direction NW? Select the correct answer **A**, **B**, **C** or **D**.

  A 045°   **B** 135°   **C** 225°   **D** 315°

2 What is the three-figure bearing for SSE? Select **A**, **B**, **C** or **D**.

  A 135°   **B** 157.5°   **C** 022.5°   **D** 112.5°

3 Find the three-figure bearing of each point from *O*.

  a
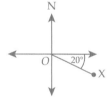
  b
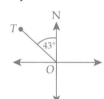
  c
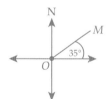

  d
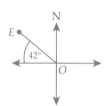
  e
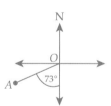
  f
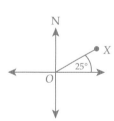

4 **a** Show the 8 main points of the compass on a compass rose.

  b Mark on your compass rose the bearings of SSW, NNE, ESE and WNW.

5 Show each set of bearings on a compass rose.

  a 020°, 295°, 110°, 232°   **b** 120°, 330°, 025°, 228°

6 Write each compass bearing as a three-figure bearing.

  a N   **b** NW   **c** E   **d** SE

  e SSE   **f** NNE   **g** WSW   **h** NNW

Nathan and Nicole sailed a boat south-east for 25 km. Nathan then stopped the boat and they had a swim. How far south (to one decimal place) are they from their starting point?

SOLUTION

Draw a diagram. Let s km be how far south they are.

$$\cos 45° = \frac{s}{25}$$
$$25\cos 45° = s$$
$$s = 17.6776\ldots$$
$$s \approx 17.7 \text{ km}$$

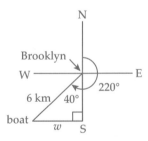

Nathan and Nicole are 17.7 km south of their starting point.

Brittany sailed from Brooklyn on a bearing of 220° for 6 km.

a   How far west (to one decimal place) was she from Brooklyn?

b   What is the bearing of Brooklyn from her boat?

SOLUTION

A bearing of 220° is 40° beyond 180°.

a   Let w km be the distance travelled west.

$$\sin 40° = \frac{w}{6}$$
$$6\sin 40° = w$$
$$w = 3.8567\ldots$$
$$w = 3.9 \text{ km}$$

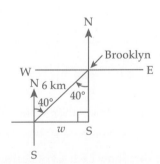

Brittany is 3.9 km west of Brooklyn.

b   When you want a bearing from a different point (the boat), draw another compass rose centred at that point.

The two marked angles are equal alternate angles on parallel lines (north–south).

The bearing of Brooklyn from the boat is 040° from north using the new compass.

1   What is the three-figure bearing of the point?
    Select the correct answer **A**, **B**, **C** or **D**.

    A 042°   **B** 138°
    C 132°   **D** 142°

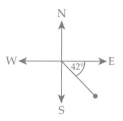

2   What is the three-figure bearing for NNW? Select **A**, **B**, **C** or **D**.
    A 022.5°   **B** 045°   **C** 315°   **D** 337.5°

3   Simon sailed 5 km northwest. How far (to one decimal place)
    is he west of his starting point?

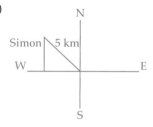

4   Jessie drove southeast from Broken Hill for 45 km. How far (to one decimal place):
    a south has she travelled?   **b** east has she travelled?

5   David chartered a yacht from a port and sailed 130° for 8 km.
    a How far is he east of the port, correct to two decimal places?
    b What is the bearing of the port from his yacht?

6   Yumi drove northeast from Perth for 150 km. How far north of Perth was she then,
    to 1 decimal place?

7   A ship is due north of Darwin. From the ship, on a bearing of 108°, a lighthouse is seen.
    If the lighthouse is 12 km due east of Darwin, how far is the ship from the lighthouse,
    to 1 decimal place?

8   A lake is 50 km due north of Botany. From the lake, a mountain due east of Botany is
    seen on a bearing of 150°. How far is the mountain east of Botany, to 1 decimal place?

Steffen Hauser / botanikfoto / Alamy

FIND-A-WORD PUZZLE

Make a copy of this puzzle, then find all the words listed below in this grid of letters.

A	R	T	R	I	A	N	G	L	E	R	A	G	I	L	Y
S	O	R	A	T	P	T	A	W	N	U	W	E	W	R	E
T	L	I	P	S	U	A	T	S	I	T	E	A	E	I	R
E	L	G	N	A	S	N	O	I	S	S	E	R	P	E	D
T	I	O	A	D	Y	I	B	N	O	E	R	T	A	L	A
I	N	N	C	J	B	O	S	E	C	A	G	A	T	E	U
S	G	O	T	A	X	A	T	R	Y	T	E	S	I	V	F
O	B	M	A	C	O	V	I	P	E	B	D	H	U	A	T
P	A	E	B	E	S	U	N	E	T	O	P	Y	H	T	R
P	T	T	A	N	A	T	O	A	R	A	R	U	Q	I	E
O	C	R	O	T	N	E	G	N	A	T	T	P	S	O	R
G	D	Y	D	B	E	R	E	T	A	S	E	T	O	N	E

ADJACENT	ANGLE	COSINE	DEGREE
DEPRESSION	ELEVATION	HYPOTENUSE	OPPOSITE
SINE	TANGENT	TRIANGLE	TRIGONOMETRY

ISBN 9780170351058

Part A General topics

Calculators are not allowed.

1   Round $235.2685 to the nearest cent.

2   Convert 45% to a simple fraction.

3   If $a = -2$, then evaluate $14 - 3a$.

4   Find the median of these scores.

1	0 1 2
2	4
3	9
4	3 6

5   What type of angle has size 180°?

6   Evaluate 75% of $48.

7   Expand $-2p(3p - 5)$.

8   Test whether (3, 2) lies on the line with equation $y = 2x + 5$.

9   How many litres in 8.4 kL?

10 Find x.

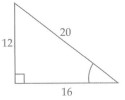

Part B Trigonometry

Calculators are allowed.

6–01 The sides of a right-angled triangle

11 For this triangle, which side is adjacent to the marked angle? Select **A**, **B**, **C** or **D**.

   A 20   **B** 16

   C 12   **D** 12 or 16

6–02 The trigonometric ratios

12 What is sin A for this triangle? Select **A**, **B**, **C** or **D**.

   A $\dfrac{9}{41}$ **B** $\dfrac{9}{40}$

   C $\dfrac{40}{9}$ **D** $\dfrac{40}{41}$

6–03 Using tan to find an unknown side

13 Find the value of each pronumeral, correct to 2 decimal places.

a

b

6–04 Using tan to find an unknown angle

14 Find the size of $\angle A$, correct to the nearest degree.

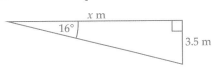

6–05 Finding an unknown side

15 Find the value of q in each triangle, correct to 1 decimal place.

a

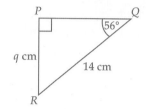

b

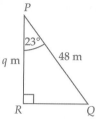

6–06 Finding more unknown sides

16 Find the value of each pronumeral, correct to 1 decimal place.

a

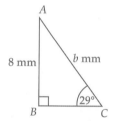

b

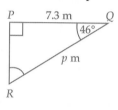

6–07 Finding an unknown angle

17 Find θ in each triangle, correct to the nearest degree.

a

b

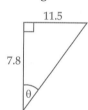

6–08 Angles of elevation and depression

18 Find the angle of elevation when looking up at a 36 m high hill from a point on the ground that is 18 m from the foot of the hill, to the nearest degree.

6–09 Bearings

19 Write each compass bearing as a three-figure bearing.

   a West   **b** NE

6–10 Problems involving bearings

20 Shaun chartered a yacht from a port and then sailed 125° from north for 12 km. How far is he east of the port, to 2 decimal places?

RATIOS AND RATES

7

IN THIS CHAPTER YOU WILL:

- find equivalent ratios and simplify ratios
- solve ratio problems
- understand how to read a scale diagram
- interpret scale maps and diagrams
- divide a quantity into a given ratio
- write and simplify rates
- solve rate problems, including those related to speed
- interpret and draw travel graphs

Shutterstock.com/mhatzapa

ISBN 9780170351058

A **ratio** compares quantities of the same kind, consisting of two or more numbers that represent parts or shares. For example, when mixing ingredients for a cake we could have a ratio of flour to milk of 3 : 2. This means there are 3 parts of flour to 2 parts of milk. The **order** is important. The ratio of milk to flour would be 2 : 3.

Operations with ratios are similar to operations with fractions.

> **To find an equivalent ratio**, multiply or divide each term by the same number.

EXAMPLE 1

Complete each pair of equivalent ratios.

a $5 : 3 = 15 :$ _____
b $12 : 30 =$ _____ $: 5$
c $3 : 5 : 4 = 6 :$ _____ $:$ _____

SOLUTION

Examine the term that has been multiplied or divided from LHS to RHS.

a $5 : 3 = 15 :$ ___ ⟵ $5 \times 3 = 15$
$5 : 3 = 15 : 9$ ⟵ Do the same to 3 to complete the ratio: $3 \times 3 = 9$

b $12 : 30 =$ ___ $: 5$ ⟵ $30 \div 6 = 5$.
$12 : 30 = 2 : 5$ ⟵ Do the same to 12 to complete the ratio: $12 \div 6 = 2$

c $3 : 5 : 4 = 6 :$ ___ $:$ ___ ⟵ $3 \times 2 = 6$
$3 : 5 : 4 = 6 : 10 : 8$ ⟵ $5 \times 2 = 10, 4 \times 2 = 8$

> **To simplify a ratio**, divide each number in the ratio by the highest common factor (HCF).

EXAMPLE 2

Simplify each ratio.

a $15 : 35$
b $75c : \$3$
c $\dfrac{1}{4} : \dfrac{2}{3}$
d $0.6 : 0.24$

SOLUTION

a $15 : 35 = \dfrac{15}{5} : \dfrac{35}{5}$ ⟵ Dividing both terms by the HCF, 5.

$= 3 : 7$

> ✳ Simplifying ratios is similar to simplifying fractions.

b $75c : \$3 = 75c : 300c$ ⟵ Convert both to cents.

$= \dfrac{75}{75} : \dfrac{300}{75}$ ⟵ Dividing both terms by the HCF, 75.

$= 1 : 4$

c $\dfrac{1}{4} : \dfrac{2}{3} = \dfrac{1}{4} \times 12 : \dfrac{2}{3} \times 12$ ⟵ Multiplying both terms by the LCD, 12, to make them whole.

$= 3 : 8$

d $0.6 : 0.24 = 0.6 \times 100 : 0.24 \times 100$ ⟵ Multiplying both terms by 100 to make them whole.

$= 60 : 24$

$= 5 : 2$ ⟵ Dividing both terms by the HCF, 12.

1 Which ratio is equivalent to 4 : 7? Select the correct answer **A**, **B**, **C** or **D**.

    A 12 : 14   **B** 8 : 21   **C** 12 : 21   **D** 16 : 35

2 Simplify 65 : 50. Select **A**, **B**, **C** or **D**.

    A 12 : 10   **B** 13 : 10   **C** 5 : 10   **D** 25 : 5

3 Copy and complete each equivalent ratio.

    a 2 : 5 = 6 : ___   **b** 5 : 3 = 30 : ___

    c 3 : 7 = ___ : 21   **d** 15 : 5 = ___ : 30

    e 14 : 60 = 7 : ___   **f** 45 : 50 = ___ : 10

    g 33 : 55 = ___ : 5   **h** 21 : 49 = 3 : ___

    i 24 : 36 = ___ : 9   **j** 81 : 72 = ___ : 8

    k 3 : 5 : 8 = 9 : ___ : ___   **l** 15 : 45 : 75 = 5 : ___ : ___

4 Simplify each ratio.

    a 12 : 18   **b** 25 : 150

    c 48 : 36   **d** 132 : 44

    e 200 : 60   **f** 20 : 48 : 60

    g $\dfrac{1}{4} : \dfrac{1}{3}$ **h** $\dfrac{2}{3} : \dfrac{1}{2}$

    i $\dfrac{3}{8} : \dfrac{1}{4}$ **j** $\dfrac{2}{5} : \dfrac{5}{6}$

    k $\dfrac{3}{2} : \dfrac{4}{5}$ **l** $\dfrac{2}{3} : \dfrac{3}{5}$

    m 0.3 : 2.4   **n** 0.5 : 4.5

    o 0.25 : 2.5   **p** 0.8 : 0.32

5 Simplify each ratio.

    a \$1.20 : \$4.80   **b** 15 cm : 60 cm

    c 2500 km : 250 km   **d** 15 s : 1 min

    e 80c : \$5   **f** 28 mm : 70 cm

    g $\dfrac{1}{4}$ cm : 1 cm   **h** $\dfrac{2}{3}$ kg : $\dfrac{1}{2}$ kg

    i 0.6 kg : 500 g   **j** 280 m : 4 km

    k 120 s : 8 min   **l** \$25 : 150c

    m 20.8 mL : 2 L   **n** 8 h : 2 days

    o 3600 g : 4.8 kg   **p** 75 min : 2 h

6 Simplify 24 : 60 : 120 : 4200.

EXAMPLE 3

There were 130 people at the cinema watching *Zombie Teachers*. 68 of them were male.

Find each ratio in simplest form.

a females : males   b males : females   c females : whole audience

SOLUTION

a Number of females = 130 − 68 = 62
   Females : males = 62 : 68
                   = 31 : 34

b Males : females = 68 : 62
                   = 34 : 31

c Females : whole audience = 62 : 130
                            = 31 : 65

EXAMPLE 4

Richard is making chocolate ice-cream by mixing milk, chocolate and cream in the ratio 3 : 2 : 1.

a How much milk and cream should be mixed with 4 cups of chocolate?

b If $2\dfrac{1}{2}$ cups of cream are used, how much chocolate is needed?

SOLUTION

Milk : chocolate : cream = 3 : 2 : 1

a 2 parts (chocolate) = 4 cups
   1 part = 4 ÷ 2   ◄——— Using the unitary method to find one part.
         = 2 cups
   3 parts (milk) = 3 × 2 = 6 cups
   6 cups of milk are required.
   1 part (cream) = 2 cups
   2 cups of cream are required

   OR: Milk : chocolate : cream = 3 : 2 : 1 = ____ : 4 : ____   ◄——— Using equivalent ratios
                            3 : 2 : 1 = 6 : 4 : 2
   6 cups of milk and 2 cups of cream are required.

Shutterstock.com/Tim UR

b 1 part (cream) = $2\dfrac{1}{2}$ cups

   2 parts (chocolate) = $2 \times 2\dfrac{1}{2}$

                     = 5 cups
   OR: Milk : chocolate : cream = 3 : 2 : 1 = ____ : ____ : $2\dfrac{1}{2}$
                            $3 : 2 : 1 = 7\dfrac{1}{2} : 5 : 2\dfrac{1}{2}$

   5 cups of chocolate are required.

1   If there are 24 000 people in the crowd watching the soccer and 15 000 are males, what is the ratio of females : males? Select the correct answer **A**, **B**, **C** or **D**.

   A 5 : 3   **B** 5 : 8   **C** 3 : 8   **D** 3 : 5

2   In Question **1**, what is the ratio of females : crowd? Select **A**, **B**, **C** or **D**.

   A 5 : 3   **B** 5 : 8   **C** 3 : 8   **D** 3 : 5

3   A class of 24 students has 12 with blue eyes, 8 with hazel eyes and 4 with brown eyes. Find the ratio of students with eye colour:

   a brown : hazel   **b** blue : brown   **c** hazel : blue

4   In a pack of animals, there are 36 sheep, 28 horses and 18 cows. Find the ratio of:

   a horses : cows   **b** sheep : cows   **c** cows : sheep

5   Ella and Dom share the rent of a house in the ratio 3 : 5. If Dom pays \$250, find Ella's share of the rent. Also find the whole rent.

6   A car dealer finds that the ratio of sedans to hatchbacks sold was 14 : 3. If 36 hatchbacks were sold, how many sedans were sold?

7   The ratio of the Tigers' wins to losses was 5 : 2. If the team won 30 games, how many games did it lose?

8   Sheldon and Raj own a shop in the ratio 4 : 3. If Sheldon's share is \$35 000, what is Raj's share?

9   Sand and cement are mixed in the ratio 4 : 1 to make concrete. How much sand is needed to mix with 32 kg of cement?

10 The ratio of wraps to salad rolls sold at a school canteen was 7 : 8. If 91 wraps were sold, how many salad rolls were sold?

11 The ratio of adults to children at the cinema was 5 : 4. If there were 100 children, how many adults were there?

12 Layne and Anna share the profits of a company in the ratio 7 : 6. If Anna receives \$468, how much does Layne get?

13 A recipe for making a cake mixes flour, sugar and milk in the ratio 5 : 2 : 3.

   a How much sugar and milk should be added to 600 mL of flour?

   b If 210 mL of milk is used, how much flour is needed?

14 A survey of car buyers found that they purchased cars with colours in this ratio:
white : red : other colours = 13 : 3 : 7

   a If a dealer ordered 156 white cars, how many red cars should he order?

   b What fraction of the cars surveyed were white?

WORDBANK

scale diagram A miniature or enlarged drawing of an actual object or building, in which lengths and distances are in the same ratio as the actual lengths and distances.

scale The ratio on a scale diagram that compares lengths on the diagram to the actual lengths.

scaled length A length on a scale diagram that represents an actual length.

A scale of 1 : 20 on a scale map or diagram means that the actual lengths are 20 times larger than on the map or diagram, whereas a scale of 20 : 1 means that the actual lengths are 20 times smaller.

The scale on a scale diagram is written as the ratio **scaled length : actual length**
- The first term of the ratio is usually smaller, meaning that the diagram is a miniature version.
- If the first term of the ratio is larger, then the diagram is an enlarged version.

EXAMPLE 5

Measure and find the actual length that each interval represents if a scale of 1 : 30 has been used.

a |————————————| b

SOLUTION

Scale is 1 : 30

a Scaled length = 5 cm
  Actual length = 5 cm × 30

   The actual length is 30 times larger

  = 150 cm
  = 1.5 m

b Scaled length = 6.9 cm
  Actual length = 6.9 cm × 30
  = 207 cm
  = 2.07 m

EXAMPLE 6

Given the scale in each diagram, measure and calculate the actual length of each object.

a

Shutterstock.com/Nature Art

Scale 2 : 1

b

Scale 1 : 4200

Shutterstock.com/Nerthuz

SOLUTION

a Scaled length = 4.8 cm
   Bee's length = 4.8 cm ÷ 2   ←——— Divide, as the scale diagram is an enlargement.
              = 2.4 cm

b Scaled length = 6.5 cm
   Ship's length = 6.5 cm × 4200   ←——— The actual ship is 4200 times larger.
              = 27 300 cm
              = 273 m

EXERCISE 7-03

1 What is the actual length of an interval if the scale length is 4 cm and the scale is 1 : 12? Select the correct answer **A**, **B**, **C** or **D**.

   A 48 cm   **B** 52 cm   **C** 3 cm   **D** $\frac{1}{3}$ cm

2 What is the length of a caterpillar if its scale length in a diagram is 8 cm and the scale is 2 : 1? Select **A**, **B**, **C** or **D**.

   A 10 cm   **B** 16 cm   **C** 4 cm   **D** 0.25 cm

3 Find the actual length (in metres) that each interval represents if a scale of 1 : 50 has been used.

4 Find the actual length of each object.

Scale 1 : 5

Scale 1 : 15

c

Scale 8 : 1

d

Scale 1 : 180

5   A spider is drawn to a scale of 4 : 1.

   a   How long are its legs if they are drawn 12 cm long?

   b   How long is its body if it is drawn 18.4 cm long?

6   Convert each scale to the same units and simplify the ratio.

   a   1 cm : 4 m   b   1 mm : 30 cm   c   3 cm : 4 km   d   5 mm : 15 cm

7   The scale used to draw this plan of a house is
    1 cm to represent 3 m.

   a   What is the actual length of the house?

   b   What is the actual width?

8   This tree is drawn to a scale of 6 mm : 1 m. What is
    the actual height of the tree in metres?

9   The scale used to represent the distance between
    Max's house (A) and Laura's house (B) is
    4 mm : 100 m.

A ⊢————————————————————————————————⊣ B

   What is the actual distance

   a   in metres?   b   in kilometres?

ISBN 9780170351058

7-04 Dividing a quantity in a given ratio

To divide a quantity in a given ratio:
- find the total number of parts by adding the terms of the ratio
- find the size of one part by dividing the quantity by the number of parts (unitary method)
- multiply to find the shares required

EXAMPLE 7

Kamil and Jonas share a lotto prize of $6417 in the ratio 5 : 4. How much do they each receive?

SOLUTION

Total number of parts = 5 + 4 = 9
One part = $6417 ÷ 9   ⟵——— Using the unitary method to find one part.
    = $713

Kamil's share = 5 × $713   ⟵——— 5 parts
    = $3565

Jonas' share = 4 × $713   ⟵——— 4 parts
    = $2852

Check: $3565 + $2852 = $6417   ⟵——— The two shares add up to the whole prize.

EXAMPLE 8

Hollee, Jesse and Jeremy sing 72 songs in the ratio 3 : 5 : 4.

How many songs do they each sing?

Stock Foundry Images / Alma

SOLUTION

Total number of parts = 3 + 5 + 4 = 12
One part = 72 ÷ 12
    = 6

Hollee's songs = 3 × 6   ⟵——— 3 parts
    = 18

Jesse's songs = 5 × 6   ⟵——— 5 parts
    = 30

Jeremy's songs = 4 × 6   ⟵——— 4 parts
    = 24

Check: 18 + 30 + 24 = 72   ⟵——— The parts add up to the whole.

1   Find the size of one part if 420 is divided in the ratio 2 : 5.
    Select the correct answer **A**, **B**, **C** or **D**.

    A 84   **B** 60   **C** 70   **D** 210

2   What does Hadieya receive if $6400 is divided between Nuraan and Hadieya in the ratio
    2 : 3? Select **A**, **B**, **C** or **D**.

    A $3840   **B** $1280   **C** $2560   **D** $3200

3   Copy and complete this table.

Ratio	Total number of parts	Total amount	One part	Ratio
2 : 1	2 + 1 = 3	45	45 ÷ 3 = 15	30 : 15
3 : 1		48		
1 : 4		300		
2 : 5		560		
4 : 3		1470		
5 : 3		24 000		

4   Divide $225 between Jasmine and Melissa in the ratio 3 : 2.

5   Divide $14 280 between Callum and Bassam in the ratio 3 : 4.

6   Amie is 14 years old and Anjit is 11 years old. They were given $5750 to be shared in the
    ratio of their ages. How much should Anjit get?

7   If $18 450 is to be shared among three people in the ratio 2 : 3 : 4,

    a   how much is the largest share?

    b   how much is the smallest share?

8   A soccer team played 35 matches in a season. If the ratio of wins to losses was 3 : 2,
    how many matches did the team lose?

9   Three friends share a prize of $156 000 in a lottery. How much does each one receive, if it is
    to be shared in the ratio 3 : 5 : 4?

10 Tori invests $25 000 and Lucy invests $35 000 in a business. If the profit at the end of the
    year is $132 000, how much should each receive if the profits are shared in the same ratio
    as their investments?

11 Divide $6446 in the ratio 5 : 2 : 4.

12 Ashleigh, Sarah and Hayley share 45 chocolates in the ratio 6 : 5 : 4. How many chocolates
    does each person receive?

ISBN 9780170351058

A **rate** compares two quantities of different kinds or different units of measure. For example, heartbeats are measured in beats/minute and the cost of petrol is measured in cents/litre. 90 beats **per** minute means 90 beats in 1 minute and is written as 90 beats/minute.

EXAMPLE 9

Write each measurement as a simplified rate.

a 360 km in 4 hours **b** $175 for 7 hours work **c** $3.72 for a 3-minute call

SOLUTION

a 360 km in 4 hours = 360 ÷ 4 km/h ⟵——— Divide by 4.
$$= 90 \text{ km/h}$$

b $175 for 7 hours = 175 ÷ 7 $/h ⟵——— Divide by 7.
$$= \$25/h$$

c $3.72 for 3 minutes = $3.72 ÷ 3 ⟵——— Divide by 3.
$$= \$1.24/\text{min}$$

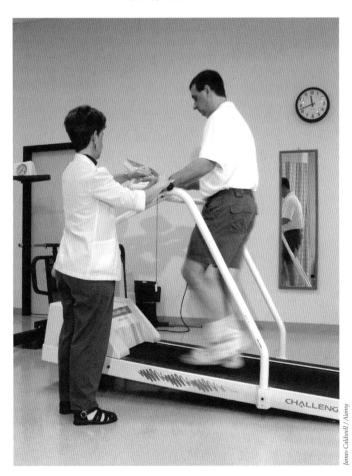

James Caldwell / Alamy

1 Simplify the rate $368 for 16 hours work. Select the correct answer **A**, **B**, **C** or **D**.

   A $22/h   **B** $22.50/h

   C $23.50/h   **D** $23/h

2 Simplify the rate 190 words in 3.8 minutes. Select **A**, **B**, **C** or **D**.

   A 50 words/min   **B** 60 words/min

   C 40 words/min   **D** 55 words/min

3 Copy and complete the table.

Rate	Simplified rate
$4.50 for 9 apples	_____ cents/apple
360 words in 8 minutes	_____ words/minute
240 km in 3 hours	_____ km/h
$1080 for 6 days at a hotel	_____ $/day
7 teachers for 168 students	_____ students/teacher
5 wickets for 125 runs	_____ runs/wicket
$18.60 for half an hour	_____ $/hour
1260 metres in 18 seconds	_____ m/s

4 An electrician took $3\frac{1}{2}$ hours to complete a job. If he charged $217, work out the rate he charged per hour.

5 A car uses 8 litres of petrol to travel 56 km. At what rate is the car using petrol, in km/L?

6 **a** Cambridge Hotel charges $660 for 4 days. What is the daily rate, in $/day?

   b Nelson Hotel charges $486 for 3 days. What is the daily rate?

   c Which hotel gives better value?

7 While a cow was browsing peacefully for 12 minutes in the grasslands, its heart beat 780 times. What was the cow's heart rate in beats/min?

8 Harry earns $152.75 for working $6\frac{1}{2}$ hours. What is his hourly rate of pay?

9 Amber's car travelled 365 km in 3.8 hours. What was her average speed in km/h, correct to 1 decimal place?

10 Which typing rate is faster: 360 words in 5 mins or 639 words in 9 mins?

ISBN 9780170351058

WORDBANK

fuel consumption The amount of fuel used by a car compared to distance travelled, measured in litres/100 km.

To solve a rate problem, write the units in the rate as a fraction $\frac{x}{y}$.
- ▪ To find x (the numerator amount), **multiply** by the rate.
- ▪ To find y (the denominator amount), **divide** by the rate.

EXAMPLE 10

Pistachio nuts cost $22.80/kg at the supermarket.

a How much does it cost to buy 0.6 kg of pistachios?

b How many kilograms of pistachios can be bought for $30.78?

SOLUTION

The units of the rate are $\dfrac{\$}{kg}$.   ⟵ Writing the units as a fraction.

a To find $, multiply by the rate.
Cost = 0.6 × $22.80   ⟵ 1 kg costs $22.80, so multiply by 0.6.
     = $13.68

b To find kg, divide by the rate.
Number of kilograms = $30.78 ÷ $22.80   ⟵ 1 kg costs $22.80, so divide by $22.80.
                = 1.35

We can buy 1.35 kg of pistachios for $30.78.

EXAMPLE 11

Lizzie took her new car for a drive to test its fuel consumption. She travelled 550 km on 68 litres of petrol. What is her fuel consumption in L/100 km, correct to two decimal places?

SOLUTION

$$\text{Fuel consumption} = \frac{68\,L}{550\ km}$$ ⟵ Writing the rate as a fraction in L/km first.

$$= \frac{68\ L}{5.5 \times 100\,km}$$ ⟵ Write 550 as a multiple of 100.

$$= \frac{68}{5.5}\ L/100\ km$$

$$= 12.3636\ldots\ L/100\ km$$

$$\approx 12.36\ L/100\ km$$

---

1   Simplify the rate 280 beats in 5 minutes in beats/min. Select the correct answer **A**, **B**, **C** or **D**.

A $\dfrac{280}{5}$ B $\dfrac{5}{280}$ C $\dfrac{1}{56}$ D   56

2   Write the fuel consumption 78 L per 480 km in L/100 km. Select **A**, **B**, **C** or **D**.

A   16.25 L/100 km   B   6.15 L/100 km   C   1.6 L/100 km   D   14 L/100 km

3   Josh earns $23 per hour working in a hardware store.

a   Write the units in the rate as a fraction.

b   How much will Josh earn if he works 28 hours?

c   How long will it take Josh to earn $368?

4   Anne types 644 words in 7 minutes.

a   How many words per minute does she type?

b   Write the units in the rate as a fraction.

c   How long will it take Anne to type 4140 words?

d   How many words will Anne type in half an hour?

5   Ellen's van travelled 300 km on 36 litres of petrol. What is its fuel consumption in L/100 km?

6   Tan paid income tax at the rate of 32c per dollar of income earned.

a   Write the units in the rate as a fraction.

b   How much tax does he pay on his income of $26 800?

c   What is his annual income if he pays $17 440 tax p.a.?

7   Find the better buy for each pair of products.

a	Betta Butter	250 g for $2.30	Marvellous Marg	400 g for $3.20
b	Jim's Jam	180 g for $2.50	June's Jam	250 g for $3.60
c	Vic's Vegemite	250 g for $3.80	Mick's Marmite	450 g for $5.20
d	Heather Honey	190 g for $3.25	Hunger Honey	350 g for $5.98
e	Cooper Coffee	450 g for $6.80	Cowper Coffee	1 kg for $10.90

8   Sam stayed 10 nights at a four-star hotel in Noosa for $1850 while Lou stayed 8 nights at a four-star hotel in Caloundra for $1600. Who found the better buy?

9   Dimitri is shopping for a new set of wheels for his car. He compared the prices of three stores: Wally's Wheels $220 each, West Wheels $780 for 4, Wonder Wheels $420 a pair. Which store offers the best buy?

10 Hand-made chocolates cost $23.50/kg.

a   Write the units in the rate as a fraction.

b   How much will 250 g of chocolates cost? Answer correct to the nearest cent.

c   How much chocolate could you buy for $25? Answer correct to the nearest gram.

11 Tegan can type at an average rate of 52 words per minute. How long, correct to the nearest minute, should it take for her to type a 6000-word essay?

12 Jonah's car travels 720 km on 85.6 L of petrol. What is his car's fuel consumption in L/100 km?

7-07 Speed

You can use this triangle to help you remember the rules.

Cover S with your finger. You are left with $\frac{D}{T}$.

If you want to find the distance, cover D and you are left with $S \times T$.

$D = S \times T$

In the same way, if you want to find time, cover T and you are left with $\frac{D}{S}$.

$T = \frac{D}{S}$

EXAMPLE 12

Chloe drove 168 km in 2 hours. What was her average speed?

SOLUTION

$S = \dfrac{168 \text{ km}}{2 \text{ h}}$ ⟵ $\dfrac{D}{T}$

$ = 84 \text{ km/h}$

EXAMPLE 13

a   If Phillip jogged at an average speed of 4 m/s, how long would he take to jog 2 km?

b   Phillip slowed down to 3 m/s and jogged for 40 minutes. How far did he go in this time?

SOLUTION

a   Rearrange the formula $S = \dfrac{D}{T}$ to find time, T.

$T = \dfrac{D}{S}$

$ = \dfrac{2000}{4}$ ⟵ $S = 4 \text{ m/s}, D = 2 \text{ km} = 2000 \text{ m}$

$ = 500 \text{ s}$

✳ **If S is in m/s, then D must be in m and T in s**

$ = (500 \div 60) \text{ min}$ ⟵ $1 \text{ min} = 60 \text{ s}$

$ = 8.333\ldots \text{ min}$

$ = 8 \text{ min } 20 \text{ s}$ ⟵ Enter ⦿ ' " or 2ndF DMS on a calculator
or calculate $0.3333\ldots \times 60$ for seconds.

b   Rearrange the formula $S = \dfrac{D}{T}$ to find the distance, D.

$D = ST$

$ = 3 \times 2400$ ⟵ $S = 3 \text{ m/s}, T = 40 \text{ min} = 40 \times 60 \text{ s} = 2400 \text{ s}$

$ = 7200 \text{ m}$

$ = 7.2 \text{ km}$

1   If it takes 4 hours to travel 340 kilometres, what is the average speed? Select the correct answer **A**, **B**, **C** or **D**.

   A 80 km/h   **B** 85 km/h   **C** 90 km/h   **D** 8.5 km/h

2   Travelling at 106 km/h, how long will it take to travel 901 km? Select **A**, **B**, **C** or **D**.

   A 8.5 h   **B** 9 h   **C** 9.5 h   **D** 7.5 h

3   Copy and complete this table to find the average speed.

Distance	Time	Speed
150 km	3 h	
750 km	5 h	
1050 km	25 h	
740 m	20 s	
4950 m	90s	

4   Brooke takes 1 hour to cycle 11 km and then walks for another half hour, travelling a further 4 km.

   a   What is the total distance travelled?

   b   How much time has she taken to travel the whole distance?

   c   Find her average speed.

5   A car travels 380 km in 4 hours.

   a   What is its average speed, in km/h?

   b   How far would this car travel in 6 hours?

   c   How long would this car take to travel 712.5 km?

6   A racing car driver does one lap of a 6.9 km race track in 3 minutes.

   a   How far would the driver travel in 40 minutes?

   b   What is the speed in km/h?

   c   How many metres would the driver travel in 60 seconds?

   d   Find his speed in m/s.

7   A car is travelling at 90 km/h.

   a   How far does it travel in one minute?

   b   How many metres is this?

   c   Find the car's speed in m/s.

8   Adrian has a practice on a race track. He does 4 laps of the 12 km track in 12 minutes.

   a   How far would he travel in 60 minutes?

   b   What is his speed in km/h?

   c   How many metres did he travel in 1 minute?

   d   Find his speed in m/s.

A **travel graph** (also called a **distance–time graph**) is a line graph that represents a journey, showing distance travelled over time. Time is shown on the horizontal axis and distance is on the vertical axis.

EXAMPLE 14

This travel graph shows Justin's trip to his friend Blake's house.

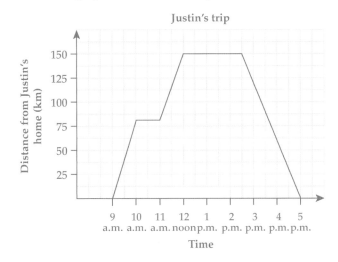

a   What time did Justin leave home?

b   For how long did he stop on the way to Blake's house?

c   When did Justin arrive at Blake's house?

d   How far is Blake's house from Justin's house?

e   How long did Justin stay at Blake's house?

f   What was Justin's average speed for the journey from Blake's house to his home?

SOLUTION

a   Justin left home at 9 a.m.   ⟵——— Graph starts at 9 a.m.

b   He stopped for 1 hour.   ⟵——— Where the line is flat between 10 and 11 a.m.

c   He arrived at 12 noon.   ⟵——— Where the line is flat again.

d   Blake's house was 150 km away.

e   Justin stayed from 12 noon to 2:30 p.m., a period of $2\frac{1}{2}$ hours.

f   For the journey home, distance travelled = 150 km, time taken = 2.5 h (from 2:30 p.m. to 5 p.m.)

$S = \dfrac{D}{T}$

$= \dfrac{150 \text{ km}}{2.5 \text{ h}}$

$= 60 \text{ km/h}$

1   What is represented on the horizontal axis of a travel graph? Select the correct answer **A**, **B**, **C** or **D**.

   A speed   **B** time   **C** metres   **D** distance

2   If part of a travel graph has a horizontal line, what does it represent? Select **A**, **B**, **C** or **D**.

   A constant speed   **B** high speed   **C** not moving   **D** a gap in time

3   This travel graph shows the bicycle trip made by a group of students during a Year 10 camp.

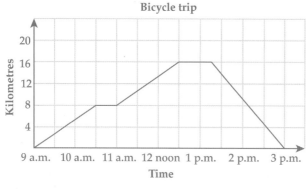

Bicycle trip

a   What time did they leave on their trip?

b   How far did they travel altogether?

c   How many times did they stop and rest?

d   What time did they start their journey back to camp?

e   Find their speed in km/h at the start of the journey, to 1 decimal place.

f   Compare this to their speed on the way back to camp, to 1 decimal place.

4 The travel graph shows a Segway tour and lunch outing.

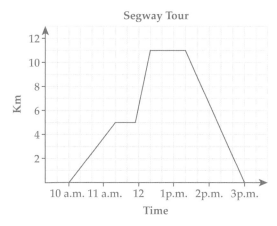

Segway Tour

a What time did they leave on their tour?
b How far did they travel altogether?
c How many times did they stop and rest?
d What time did they start their journey home?
e Find their average speed in km/h for their trip on the way out.
f Compare this to their speed on the way back home.

5 Write a story about the travel graph below. Include as many details as you can.

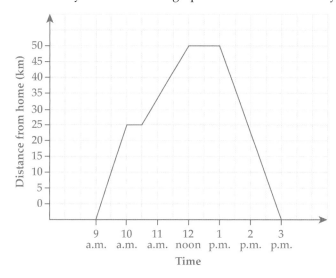

COMPOUND WORDS

The words in the list below are called **compound words** as they are made up of two smaller words. For each compound word, identify the two smaller words, count the number of letters in each word, and write the pair of numbers as a simplified ratio.

For example, SEABREEZE = SEA + BREEZE, with ratio 3 : 6 = 1 : 2.

1   LIFETIME

2   SOMETHING

3   WITHOUT

4   EARTHQUAKE

5   BASKETBALL

6   UPSTREAM

7   FIREWORKS

8   PEPPERMINT

9   AIRPORT

10 SKATEBOARD

11 SUNFLOWER

12 SOUTHWEST

13 THUNDERSTORM

14 WIDESPREAD

15 AFTERNOON

16 HEADQUARTERS

Part A General topics

Calculators are not allowed.

1   Convert 3.25% to a decimal.

2   Find the highest common factor of
    $2bc^2$ and $10ab$.

3   Find the value of a,
    correct to one
    decimal place.

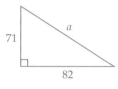

4   Evaluate $\dfrac{2}{5} \times \dfrac{25}{32}$

5   Evaluate $(-2)^6$

6   Simplify $xy - px + 5xy + xp$.

7   Find θ, correct to the nearest degree.

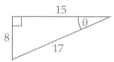

8   How much simple interest is earned if
    $8000 is invested at 5% p.a. for 3 years?

9   A die is rolled. What is the probability of
    rolling a factor of 6?

10 How many months are there in 3 years?

Part B Ratios and rates

Calculators are allowed.

7–01 Ratios

11 Simplify the ratio 180 : 54. Select the correct answer **A, B, C** or **D**.

    A 20 : 6   **B** 90 : 27   **C** 18 : 5.4   **D** 10 : 3

12 Which of the following is an equivalent ratio for 1 : 2 : 5? Select **A, B, C** or **D**.

    A 3 : 6 : 12   **B** 5 : 10 : 25   **C** 4 : 12 : 20   **D** 7 : 14 : 28

7–02 Ratio problems

13 At a conference for 64 adults, there are 28 males. What is the ratio of females : adults?
    Select **A, B, C** or **D**.

    A 9 : 16   **B** 36 : 28   **C** 28 : 64   **D** 64 : 36

14 Corrina mixes juice using 3 parts apples, 2 parts orange and 0.5 part watermelon.
    Find each ratio in simplified form.

    a   orange : apple : watermelon   b   orange : whole juice

7–03 Scale maps and diagrams

15 Find the actual length of each figure.

    a   b

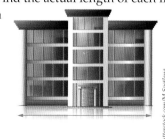

Shutterstock.com/M.Svetlana

Shutterstock.com/VooDoo13

Scale 1 : 20

Scale 1 : 300

7–04 Dividing a quantity in a given ratio

16 Divide a prize of $72 000 between Ray, Ed and Pete in the ratio 1 : 2 : 3.

7–05 Rates

17 Simplify each rate.

    a $210 for 3.5 hours work   **b** $7.80 for 12 apples   **c** 390 words in 5 minutes

18 During a run, Jade's heart beat 423 times in $4\frac{1}{2}$ minutes. What was her heart rate?

7–06 Rate problems

19 In the supermarket, beef costs $24.60/kg.

    a How much will it cost to buy 250 g of beef?

    b How many grams of beef could I buy for $17.70?

7–07 Speed

20 A bus is travelling at 75 km/h.

    a How far does it travel in 1 minute?

    b How many metres is this?

    c Find the bus' speed in m/s, to 1 decimal place.

7–08 Travel graphs

21 What is shown on the vertical axis of a travel graph?

22 **a** For this travel graph, find the total distance travelled during the trip.

    b When did the person stop?

    c Calculate correct to one decimal place the speed during the start of the trip.

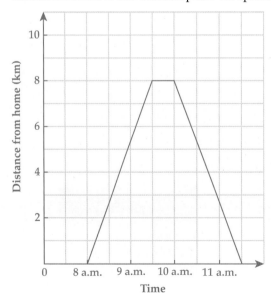

CONGRUENT AND SIMILAR FIGURES

8

IN THIS CHAPTER YOU WILL:

- name and classify triangles and their properties
- name and classify quadrilaterals and their side, angle and diagonal properties
- solve geometry problems involving the properties and angle sums of triangles and quadrilaterals
- perform transformations: translate, reflect and rotate shapes
- identify congruent figures and their properties
- use the '≡' symbol
- identify the four tests for congruent triangles: SSS, SAS, AAS, RHS
- identify similar figures and their properties
- use scale factors to enlarge or reduce a shape
- find the length of unknown sides in similar figures

Shutterstock.com/Ratikova

Classifying triangles by sides

Equilateral triangle	Isosceles triangle	Scalene triangle
All sides equal	Two sides equal	All sides different
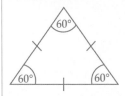		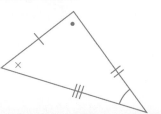

Classifying triangles by angles

Acute-angled triangle	Obtuse-angled triangle	Right-angled triangle
All angles acute	One angle obtuse	One right angle
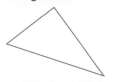		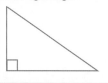

The **angle sum of any triangle** is 180°.

$a + b + c = 180$

The **exterior angle of a triangle** is the sum of the two interior opposite angles.

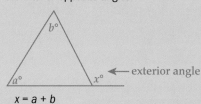

$x = a + b$

EXAMPLE 1

Find the value of each pronumeral, giving reasons.

a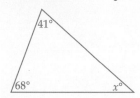

b

c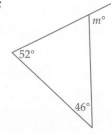

SOLUTION

a $68 + 41 + x = 180$ (angle sum of a triangle)
$$x = 180 - 68 - 41$$
$$x = 71$$

b $44 + 44 + n = 180$ (angle sum of an isosceles triangle)
$$n = 180 - 44 - 44$$
$$n = 92$$

c $m = 52 + 46$ (exterior angle of a triangle)
$$m = 98$$

1   What type of triangle has three equal sides? Select the correct answer **A**, **B**, **C** or **D**.

   A equilateral   **B** right-angled   **C** scalene   **D** isosceles

2   In an isosceles triangle, where are the equal angles? Select **A**, **B**, **C** or **D**.

   A All three angles are equal   **B** At the bottom of the triangle
   C At the top of the triangle   **D** Opposite the equal sides

3   Classify each triangle according to its sides.

   a   **b**    **c**

4   Classify each triangle in Question **3** according to its angles.

5   Find the value of each pronumeral.

   a   **b** 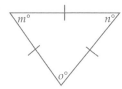   **c**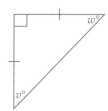

6   For each triangle, name:

   i   the exterior angle   **ii**   the two interior opposite angles

   a 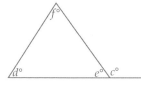   **b**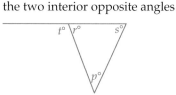

7   Find the value of each pronumeral, giving reasons.

   a    **b** 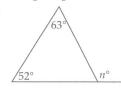   **c**

   d    **e**    **f**

8   Sketch each triangle described.

   a   Isosceles and right-angled   **b**   Scalene and obtuse-angled

Convex quadrilateral	Non-convex quadrilateral	Parallelogram
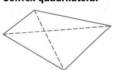 All vertices point outwards. All diagonals lie inside the shape.	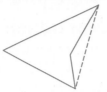 One vertex points inwards. One diagonal lies outside the shape.	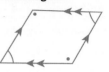
Trapezium 	Rectangle 	Rhombus
Square 	Kite 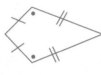	

The **angle sum of any quadrilateral** is 360°.
$a + b + c + d = 360$

EXAMPLE 2

Find the value of each pronumeral, giving reasons.

a 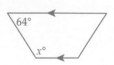   b    c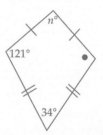

SOLUTION

a $x + 64 = 180$ (co-interior angles on parallel lines)
       $x = 180 - 64$
       $x = 116$

b $v = 114$ (opposite angles of a parallelogram)

c   • = 121 (opposite angles of a kite)
     $121 + 121 + 34 + n = 360$ (angle sum of a quadrilateral)
     $n = 360 - 121 - 121 - 34$
     $n = 84$

ISBN 9780170351058

1   Which quadrilateral has equal diagonals? Select the correct answer **A, B, C** or **D**.
    A rhombus   **B** rectangle   **C** parallelogram   **D** trapezium

2   What is the size of all angles in a square? Select **A, B, C** or **D**.
    A 180°   **B** 90°   **C** 360°   **D** 60°

3   Draw a sketch of each quadrilateral.
    a trapezium   **b** kite   **c** non-convex quadrilateral
    d square   **e** convex quadrilateral **f** parallelogram

4   Name each quadrilateral.

a

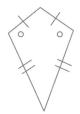

b

c

5   Find the value of each pronumeral, giving reasons.

a

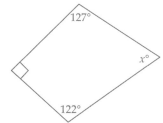

127°
$x°$
122°

b

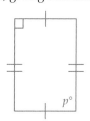

$p°$

c

93°
$n°$
48°

d

62°
$x°$

e

124°
$v°$

f

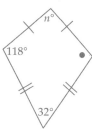

$n°$
118°
32°

g

$n°$

h

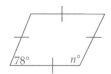

78°
$n°$

i

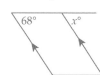

68°
$x°$

WORDBANK

transformation The process of moving or changing a shape, by translation, reflection or rotation.

translation The process of 'sliding' a shape, moving it up, down, left or right.

reflection The process of 'flipping' a shape, making it back-to-front like in a mirror.

rotation The process of 'spinning' a shape, around a point, tilting it sideways or upside-down

EXAMPLE 3

a This L-shape has been translated 3 units to the right.

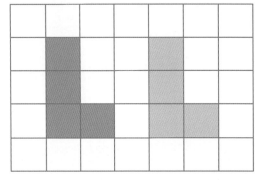

b This L-shape has been reflected across the line *AB*.

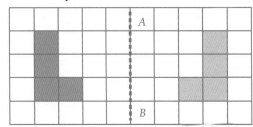

c This L-shape has been rotated 90° clockwise about point *O*.

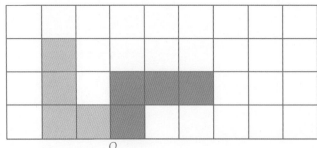

1   What type of transformation occurs if a shape is turned anticlockwise? Select the correct answer **A, B, C** or **D**.

   A   translation   **B**   reflection   **C**   transformation **D**   rotation

2   What type of transformation occurs if a shape is flipped over a line? Select **A, B, C** or **D**.

   A   translation   **B**   reflection   **C**   transformation **D**   rotation

3   Copy this rhombus on grid paper and perform each translation to it.

   a   2 units right   **b**   2 units up   **c**   1 unit left   **d**   1 unit down

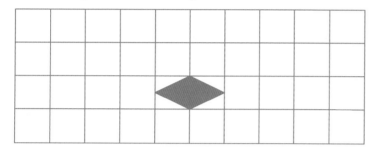

4   Copy each shape on grid paper in the line drawn.

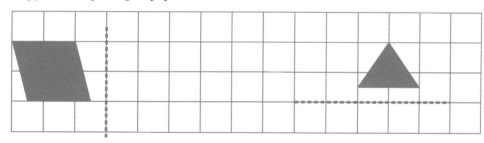

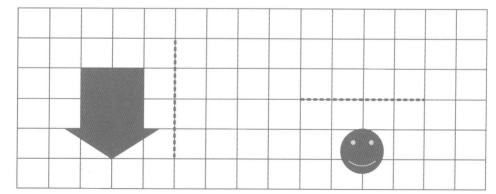

5   Copy this diagram onto grid paper and perform each rotation.

   a   90° clockwise about *O*   **b**   180° about *P*

   c   90° anticlockwise about *P*   **d**   270° clockwise about *O*

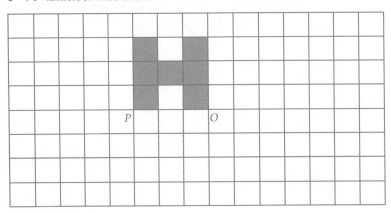

6   Copy each diagram onto grid paper and perform each rotation about point *O*.

   a   **b**

7   Copy each shape onto square dot paper and translate it in the direction and distance
   indicated by the arrow.

   a   **b**

 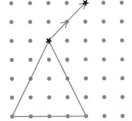

Congruent figures are shapes that are identical in every way, with the same shape and the same size.
- Matching sides are equal
- Matching angles are equal

These two triangles are congruent.

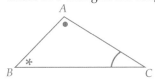 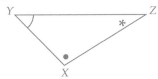

$\angle A$ matches $\angle X$, $\angle B$ matches $\angle Z$, $\angle C$ matches $\angle Y$.
We can write '$\triangle ABC \equiv \triangle XZY$' in matching order of the vertices, where '$\equiv$' stands for 'is congruent to'.
Congruent figures can be formed by **translation**, **reflection**, **rotation** or a combination of them.

EXAMPLE 4

These two triangles are congruent.

 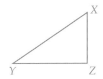

a Which transformation turns the first triangle into the second triangle?

b List the three pairs of matching angles.

c Copy and complete this congruence statement: $\triangle ABC \equiv \triangle$ ___

SOLUTION

a Reflection

b $\angle A = \angle X$, $\angle B = \angle Z$, $\angle C = \angle Y$

c $\triangle ABC \equiv \triangle XZY$ ⟵ matching order of vertices

1 If these two quadrilaterals are congruent, which angle in *PQRS* matches ∠*A*? Select the correct answer **A**, **B**, **C** or **D**.

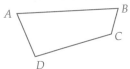 

   A *P*   **B** *Q*   **C** *R*   **D** *S*

2 For Question **1**, which is the correct order of vertices to complete the congruence statement '*ABCD* ≡ _____'? Select **A**, **B**, **C** or **D**.

   A *PQRS*   **B** *QSRP*   **C** *RSPQ*   **D** *SRQP*

3 Find three pairs of congruent figures and state which transformation can be used between them.

   A   **B**   **C**   **D**

   E   **F**   **G**   **H**

4 If △*ABC* ≡ △*RPQ*, list all pairs of matching sides and matching angles.

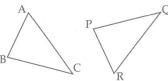

5 These two pentagons are congruent.

   a List all pairs of matching angles.

   b Copy and complete: *JKLMN* ≡ _____.

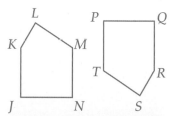

6 The Dutch artist Maurits Cornelius Escher used congruent shapes to draw amazing works of art. Describe the congruent shapes in this Escher drawing. Which transformation has he used?

Peter van Exert / Alamy

Tests for congruent triangles

There are 4 tests to prove that two triangles are congruent.

1. SSS Three sides of one triangle are equal to three sides of the other triangle.

2. SAS Two sides and the **included** angle of one triangle equal two sides and the **included angle** of the other triangle.

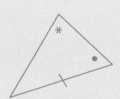 

✳ The included angle means the angle in between the two sides.

3. AAS Two angles and one side of one triangle equal two angles and the matching side of the other triangle.

4. RHS A right angle, hypotenuse and a side of one triangle equal to the right angle, hypotenuse and the matching side of the other triangle.

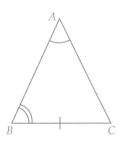

EXAMPLE 5

Which two triangles are congruent? Which congruence test proves this?

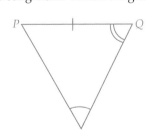 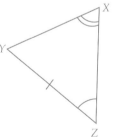

SOLUTION

$\triangle ABC \equiv \triangle RQP$ Test used was AAS.

$\triangle XYZ$ is not congruent to these triangles because YZ is not the matching side.

1   Which test proves that these triangles are congruent? Select the correct answer **A**, **B**, **C** or **D**.

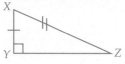 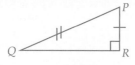

   A SAS   **B** RHS   **C** SSS   **D** AAS

2   For Question **1**, $\triangle XYZ \equiv$ _____? Select **A**, **B**, **C** or **D**.

   A $\triangle PQR$ **B** $\triangle RQP$ **C** $\triangle RPQ$ **D** $\triangle PRQ$

3   Find three pairs of congruent triangles and state which test you used.

A 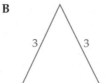   **B** 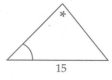   **C**

D   **E**   **F**

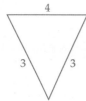 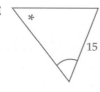

4   Draw two congruent triangles that follow:

   a the SAS test   **b** the RHS test   **c** the AAS test

5   Find three pairs of congruent triangles. Remember to write the vertices in matching order and state which test you used.

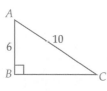 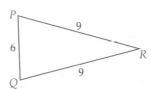 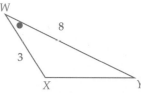

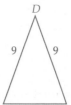 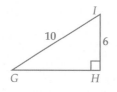 

Similar figures have the same shape but not necessarily the same size. One is an enlargement (or reduction) of the other.
- Matching angles are equal
- Matching sides have lengths that are in the same ratio

These two quadrilaterals are similar.

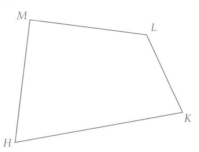

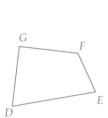

$\angle G = \angle M$, $\angle F = \angle L$, $\angle E = \angle K$, $\angle D = \angle H$.
Each side of *MLKH* is exactly double the length of the matching side in *GFED*.
We can write '*GFED* ||| *MLKH*' in matching order of the vertices, where '|||' stands for 'is similar to'.

EXAMPLE 6

Find pairs of similar figures.

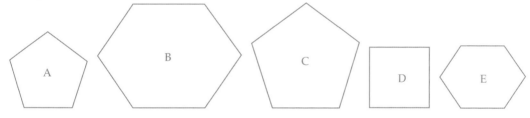

SOLUTION

A and C, B and E. ←———— They are the same shape but different sizes.

1   Which phrase is correct about similar figures? Select the correct answer **A, B, C** or **D.**

    A same shape, same size   **B** same shape, different size

    C different shape, same size   **D** different shape, different size

2   Which pair of shapes are always similar? Select **A, B, C** or **D.**

    A any two triangles   **B** any two rectangles

    C any two parallelograms   **D** any two squares

3   Which set of shapes include similar figures?

    A    **B**

    C

    D    **E**

4   Find 7 pairs of similar figures.

5 Name two pairs of similar triangles. Remember to write the vertices in matching order.

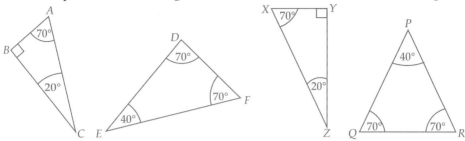

6 Two pairs of matching angles are equal in these two triangles.

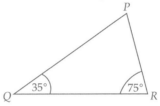

a Find the size of ∠A and ∠P.

b Is angle A the same size as angle P?

c Is △ABC ||| △PQR?

7

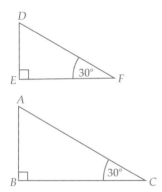

a What are the sizes of angles A and D?

b Are the two triangles similar?

c Name pairs of matching sides.

d Complete: △DEF ||| _____.

WORDBANK

enlargement A shape or diagram that has been made larger in size.

reduction A shape or diagram that has been made smaller in size.

image A transformed shape or diagram after it has been enlarged or reduced.

scale factor The amount the original figure is multiplied by to create an enlargement or reduction.

$$\text{Scale factor} = \frac{\text{image length}}{\text{original length}}$$

- If the scale factor is greater than 1, then the image is an enlargement.
- If the scale factor is between 0 and 1, then the image is a reduction.

EXAMPLE 7

Find the scale factor for each enlargement or reduction.

a

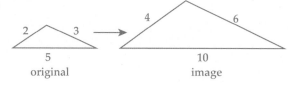

5
original

10
image

b

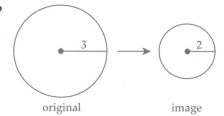

original   image

SOLUTION

a Scale factor = $\dfrac{4}{2}$ (or $\dfrac{6}{3}$ or $\dfrac{10}{5}$)   ←——— $\dfrac{\text{image length}}{\text{original length}}$

= 2

The image is double the size of the original: an enlargement.

b Scale factor = $\dfrac{2}{3}$ ←——— $\dfrac{\text{image length}}{\text{original length}}$

The image is $\dfrac{2}{3}$ the size of the original: a reduction.

1   If the original figure had a height of 4 cm and the image is 3 cm high, what is the scale factor? Select the correct answer **A**, **B**, **C** or **D**.

   A $\dfrac{4}{3}$ **B** 12   **C** $\dfrac{3}{4}$ **D** 1.3

2   In an enlargement with scale factor 1.5, the original figure is 12 cm long. What is the length of the image? Select **A**, **B**, **C** or **D**.

   A 8 cm   **B** 16 cm   **C** 24 cm   **D** 18 cm

3   Find the scale factor for each enlargement or reduction.

   a

   b

   c

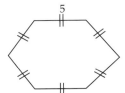

   d
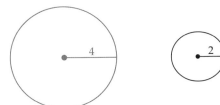

   e

   f

g

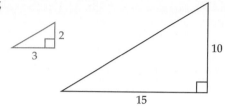

h

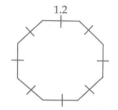

4 Enlarge each figure using a scale factor of:

   i 2   **ii** 1.5

  a    **b**   **c**

5 Reduce each figure using a scale factor of:

   i $\dfrac{1}{2}$ **ii** 0.8

  a   **b**   **c**

6 a Enlarge $\triangle ABC$ using a scale factor of 2.5.

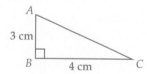

  b Reduce $\triangle ABC$ using a scale factor of $\dfrac{2}{3}$.

EXAMPLE 8

Find the value of each pronumeral in this pair of similar triangles.

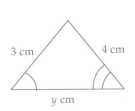

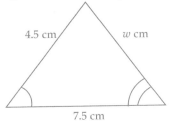

SOLUTION

First calculate the scale factor by comparing matching sides 3 cm and 4.5 cm.

Scale factor $= \dfrac{4.5}{3}$

$= 1.5$

$\therefore w = 4 \times 1.5$ ←—— 4 and w are matching sides.

$= 6$

> ✱ From the diagram, a length of 6 cm looks reasonable for w.

$\therefore y = 7.5 \div 1.5$ ←—— y and 7.5 are matching sides.

$= 5$

> ✱ From the diagram, a length of 5 cm looks reasonable for y.

1   Which phrase is true about matching sides in similar figures? Select the correct answer **A, B, C** or **D**.

    A equal
                 B twice the original

    C in the same ratio
             D random

2   Which phrase is true about matching angles in similar figures? Select **A, B, C** or **D**.

    A equal
                 B twice the original

    C in the same ratio
             D random

3   **a**   Copy and complete this table of measurements for the diagram.

	Length	Height	Door length	Door height
Small house	4 units			
Big house				
Scale factor				

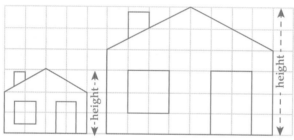

    b   Are the houses similar figures?

4   Find the value of the pronumeral in each pair of similar figures.

    a
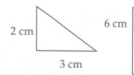

    b
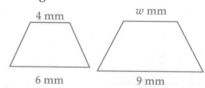

    c

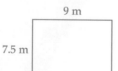

    d
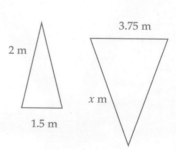

5   In this diagram, $\angle A = \angle D$ and $\triangle ABC \,|||\, \triangle DEC$. Find the length of AC.

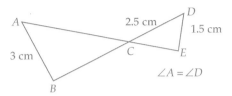

6   **a**   Copy and complete: $\triangle ABC \,|||\,$ _____.

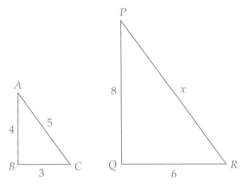

   b   Show that $\triangle ABC$ is a right-angled triangle.

   c   Find the value of x using:

   i   similar figures   **ii**   Pythagoras' theorem

Shutterstock.com/DT10

WORD SCRAMBLE

Unscramble the letters below to form words used in this chapter.

1 IMISARL
2 GLANTIRE
3 TRONCUGEN
4 GERIFU
5 SLANTRITOAN
6 TREDICOUN
7 TALAQUILRIDAR
8 GERMLENAENT

9 FLEIONTECR
10 MATAFORSNORNIT
11 SETT
12 TORNAITO
13 AMIGE
14 GLANE
15 TIMGCAHN

PRACTICE TEST 8

Part A General topics

Calculators are not allowed.

1 Evaluate $\dfrac{2}{3} - \dfrac{1}{5}$.

2 Factorise $6xy^2 + 2y$.

3 Find the mean of 1, 2, 8, 3, 6.

4 Ruchi pays a grocery bill of $83.45 with a $100 note. Calculate the change.

5 Use three letters to name the marked angle.

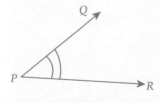

6 What type of angle is marked in Question 5?

7 Write Pythagoras' theorem for this triangle.

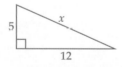

8 Find the value of x in Question 7.

9 Simplify $\dfrac{12a^2b}{3b^2}$.

10 Write a simplified algebraic expression for the perimeter of this rectangle.

Part B Congruent and similar figures

Calculators are allowed.

8-01 Triangles

11 What type of triangle is this? Select the correct answer **A, B, C** or **D**.

    A acute-angled and isosceles   **B** acute-angled and scalene
    C right-angled and isosceles   **D** right-angled and scalene

12 Find the value of x. Select **A, B, C** or **D**.

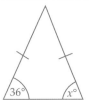

    A 72°   **B** 36°   **C** 108°   **D** 180°

8-02 Quadrilaterals

13 Name each quadrilateral.

    a 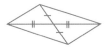   **b**

14 Find the value of each pronumeral, giving reasons.

    a    **b**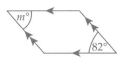

8-03 Transformations

15 Name the type of transformation that has been performed on quadrilateral *ABCD* to create the congruent figure on the right.

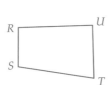

8-04 Congruent figures

16 Copy and complete for the quadrilaterals in Question 15: $ABCD \equiv$ _____

8-05 Tests for congruent triangles

17 Which test proves that these triangles are congruent?

8-06 Similar figures

18 Which two figures are similar?

A   **B**   **C**   **D**

8-07 Enlargements and reductions

19 Find the scale factor for each pair of similar figures.

a
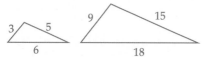

b
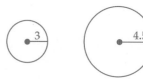

8-08 Finding sides in similar figures

20 Find the value of *x* in these similar figures.

LENGTH AND TIME

9

IN THIS CHAPTER
YOU WILL:

- calculate the perimeter of shapes, including circles and composite shapes
- calculate with time, including 24-hour time
- read timetables, understand and use international time zones, and solve problems involving them

Shutterstock.com/WitR

ISBN 9780170351058

9–01 Perimeter

To find the **perimeter** of a shape, add the lengths of its sides.

EXAMPLE 1

Find the perimeter of each shape.

a
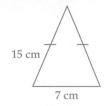

b
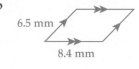

c
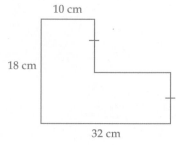

SOLUTION

a   Perimeter = 15 + 15 + 7   ⟵────── 2 equal sides in the isosceles triangle
             = 37 cm

b   Perimeter = 2 × 6.5 + 2 × 8.4   ⟵────── opposite sides are equal in the parallelogram
             = 29.8 mm

c   First find the lengths of the unknown sides.
    Perimeter = 10 + 9 + 22 + 9 + 32 + 18
             = 100 cm

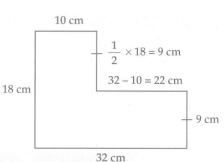

ISBN 9780170351058

1   What is the perimeter of a square of side length 8 cm? Select the correct answer **A, B, C** or **D**.
   A 64 cm   **B** 16 cm   **C** 32 cm   **D** 24 cm

2   What is the perimeter of a rectangle with length 12 mm and width 9 mm? Select **A, B, C** or **D**.
   A 108 mm   **B** 34 mm   **C** 21 mm   **D** 42 mm

3   Find the perimeter of each rectangle.
   a

   4 cm
   15 cm

   b
   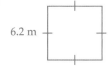
   6.2 m

   c

   3.8 cm
   9.65 cm

4   Copy and complete this table.

Shape	Side lengths	Perimeter
Rectangle	7.5 cm and 3 cm	
Square	13 mm	
Equilateral triangle	5.4 cm	
Rhombus	7.2 m	
Parallelogram	8 cm and 6 cm	
Kite	4.6 cm and 7.8 cm	

5   Find the perimeter of each shape. All measurements are in cm.

   a
   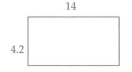
   14
   4.2

   b

   11
   15.6

   c
   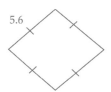
   5.6

   d
   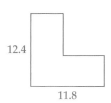
   12.4
   11.8

   e
   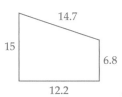
   14.7
   15
   6.8
   12.2

   f
   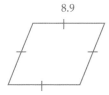
   8.9

6 Calculate the perimeter of each shape.

a

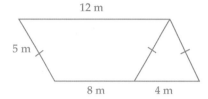

12 m

5 m

8 m   4 m

b

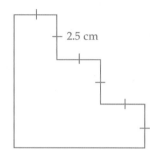

2.5 cm

c

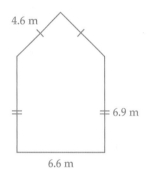

4.6 m

6.9 m

6.6 m

d

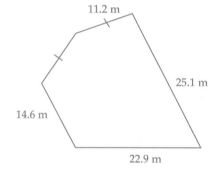

11.2 m

25.1 m

14.6 m

22.9 m

7 Josh wants to build a fence around his rectangular garden that measures 3.6 m by 1.8 m.

    a How many metres of fencing will Josh need?

    b What will be the cost of the fence if he needs 10 palings per metre and each paling costs $3.20 each?

Radius	Diameter	Circumference
The distance from the centre of a circle to the edge. The plural of radius is **radii**.	The distance from one side of a circle to another, through the centre.	The perimeter of a circle.

Arc	Sector	Quadrant
Part of the circumference of a circle.	A fraction of a circle cut along two radii.	A quarter of a circle.

		Semicircle
		Half of a circle.

Chord	Segment	Tangent
The distance from one side of a circle to another but not necessarily through the centre.	A region of a circle cut off by a chord.	A line outside a circle that touches it at exactly one point.

1   The radius of a circle is the distance between what? Select the correct answer **A**, **B**, **C** or **D**.

   A   any two points on a circle

   B   the centre and the edge of a circle

   C   two points and the centre of a circle

   D   the centre and a chord of a circle

2   The diameter of a circle is the distance between what? Select **A**, **B**, **C** or **D**.

   A   any two points on a circle

   B   the centre and the edge of a circle

   C   the edges of a circle through the centre

   D   the centre and a chord of a circle

3   Write down the name of each part of the circles shown below.

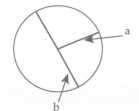   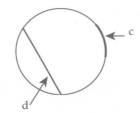

4   Using a pair of compasses, draw a circle of radius 4 cm.

   a   Mark in the centre of the circle, label it *O*.

   b   Draw in the radius, label it *OA*.

   c   Draw in a diameter and label it *CD*.

   d   Measure *OA* and *CD*.

   e   What is the connection between *OA* and *CD*?

   f   What is the connection between the radius and the diameter of a circle?

5   Name each part of the circle labelled below.

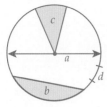   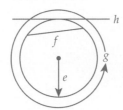

6 Name the part of the circle described.

   a A quarter of a circle.

   b A line that touches the outside of the circle once.

   c The distance from the centre of the circle to its edge.

   d A fraction of the circle's circumference.

   e The shape of a pizza slice.

   f An interval from one edge of a circle to another edge, not through the centre.

   g The area inside a circle between a chord and an arc.

   h The distance around a circle.

   i Half of a circle.

   j The distance from one edge of a circle to another edge, through the centre.

7 Is each statement true or false?

   a A quadrant has half the area of a semicircle.

   b A segment has half the area of a sector.

   c The radius is half as long as the diameter.

   d A tangent joins two points on the circle.

   e A semicircle is a special type of sector.

8 Name each part of the circle drawn.

   a   b

   c   d

   e   f

9 Use compasses, ruler and pencil to copy each design.

   a 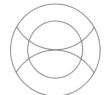   b    c

Circumference of a circle

To find the **circumference of a circle**, multiply π by the circle's diameter.

$C = \pi d$ means $C = \pi \times$ diameter

or $C = 2\pi r$ means $C = 2 \times \pi \times$ radius

EXAMPLE 2

Find the perimeter of each figure, correct to 1 decimal place.

a
7.5 m

b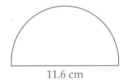
11.6 cm

c
9 m
135°

SOLUTION

a $C = 2\pi r$ ←— radius = 7.5

$= 2 \times \pi \times 7.5$ ←— 2 $\boxed{\times}$ $\boxed{\pi}$ $\boxed{\times}$ 7.5 $\boxed{=}$

$= 47.1238\ldots$

≈ 47.1 m

b Perimeter $= \dfrac{1}{2}\pi d + 11.6$ ←— diameter = 11.6

$= \dfrac{1}{2}\pi \times 11.6 + 11.6$

$= 29.8212\ldots$

≈ 29.8 cm

c Perimeter $= \dfrac{135}{360} \times 2\pi r + 9 + 9$ ←— radius = 9

✱ There are 360° in a circle, but a sector is a fraction of a circle

$= \dfrac{135}{360} \times 2 \times \pi \times 9 + 18$

$= 39.2057\ldots$

$= 39.2$ m

EXAMPLE 3

Find the perimeter of this shape:

a correct to two decimal places

b in terms of π.

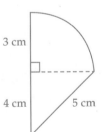
3 cm
4 cm
5 cm

SOLUTION

a Perimeter $= \dfrac{1}{4} \times 2\pi r + 3 + 4 + 5$ ←— $\dfrac{1}{4} \times$ circumference + 3 sides

$= \dfrac{1}{4} \times 2\pi \times 3 + 12$

$= 16.7123\ldots$

≈ 16.71 cm

b Perimeter $= \dfrac{1}{4} \times 2\pi \times 3 + 12$ ←— from part **a**

$= \left(\dfrac{3\pi}{2} + 12 \right)$ cm ←— in terms of π

ISBN 9780170351058

1 Find the circumference of a circle with diameter 8 cm. Select the correct answer **A, B, C** or **D**.

   A 12.57 cm   **B** 50.27 cm   **C** 25.13 cm   **D** 201.06 cm

2 Find the circumference of a circle with radius 4.2 m. Select **A, B, C** or **D**.

   A 13.2 m   **B** 26.4 m   **C** 52.8 m   **D** 55.4 m

3 Find the perimeter of each shape, correct to one decimal place.

   a

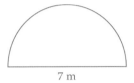

7 m

   b

4 cm

   c

4.2 m

   d

9..3 m

   e

16 m   40°

   f

120°

15 mm

4 Write your answers to Question **3 a** and **b** in terms of π.

5 Find the perimeter of each shape, correct to 1 decimal place.

   a

9.6 m

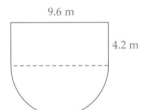

4.2 m

   b

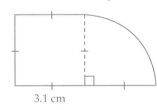

3.1 cm

   c

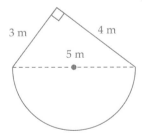

3 m   4 m   5 m

   d

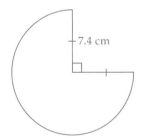

7.4 cm

6 **a** Jake wants to surround his semicircular garden with a picket fence. The radius of the garden is 3.8 m. How long (correct to one decimal place) does the fence need to be?

   b Jesse buys the palings for the fence in Jake's garden and needs 10 palings for each metre of fencing. How much does it cost her for the palings if they are $4.20 each?

24-hour time uses 4 digits to describe the time of day and does not require a.m. or p.m. 0000 means 12 midnight, 0100 means 1 a.m., 1200 means 12 midday, 1300 means 1 p.m. and so on, up till 2300, which means 11 p.m.

EXAMPLE 4

Write each time in 24-hour time.

a 3:24 a.m.   **b** 7:16 p.m.   **c** 12.48 a.m.

SOLUTION

a 3:24 a.m. = 0324 in 24-hour time ←——— After 1 a.m., insert a 0 in front to make 4 digits.

b 7:16 p.m. = 1916 in 24-hour time ←——— After 1 p.m., add 12 to the hour: 7 + 12 = 19

c 12.48 a.m. = 0048 in 24-hour time ←——— 12 midnight is 00 for first 2 digits.

EXAMPLE 5

Round 42 min 38 s to the nearest minute.

SOLUTION

42 min 38 s ≈ 43 min.   ←——— 36 s is more than 30 s (half a minute), so round up.

EXAMPLE 6

Convert:

a 132 minutes to hours and minutes   **b** 6.3 hours to hours and minutes

SOLUTION

a 132 minutes = (132 ÷ 60) h   ←——— 1 h = 60 min
$$= 2.2 \text{ h}$$
$$= 2 \text{ h } 12 \text{ min}$$ ←——— Enter $\boxed{\circ\,'\,''}$ or $\boxed{\text{2ndF}}$ $\boxed{\text{DMS}}$ on a calculator
or calculate 0.2 × 60 for minutes

b 6.3 hours = 6 hours + 0.3 × 60 min   ←——— or enter 6.3 $\boxed{=}$ $\boxed{\circ\,'\,''}$ or $\boxed{\text{2ndF}}$ $\boxed{\text{DMS}}$ on a calculator
$$= 6 \text{ hours } 18 \text{ min}$$

EXAMPLE 7

Calculate the time difference between 4:40 p.m. and 10:50 p.m.

SOLUTION

Use a number line and 'build bridges' like we do for mental subtraction.

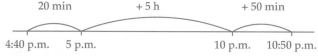

Time difference = 20 min + 5 h + 50 min
$$= 5 \text{ h} + 70 \text{ min}$$
$$= 5 \text{ h} + 1 \text{ h} + 10 \text{ min}$$
$$= 6 \text{ h} + 10 \text{ min}$$

OR: convert to 24-hour time and use the calculator's $\boxed{\circ\,'\,''}$ or $\boxed{\text{DMS}}$ keys.

4:40 p.m. = 1640, 10:50 p.m. = 2250, so enter 22 $\boxed{\circ\,'\,''}$ 50 $\boxed{\circ\,'\,''}$ $\boxed{-}$ 16 $\boxed{\circ\,'\,''}$ 40 $\boxed{\circ\,'\,''}$ $\boxed{=}$

1   Convert 6:42 p.m. to 24-hour time. Select the correct answer **A, B, C** or **D**.
    A 1642   **B** 0642   **C** 1242   **D** 1842

2   Convert 8.4 hours to hours and minutes. Select **A, B, C** or **D**.
    A 8 h 4 min   **B** 8 h 24 min   **C** 8 h 32 min   **D** 8 h 40 min

3   Convert each time to 24-hour time.
    a   3:50 a.m.   **b**   4:28 p.m.   **c**   5:36 a.m.   **d**   8:45 p.m.
    e   12 midnight **f**   9:25 a.m.   **g**   12:54 a.m.   **h**   11:55 p.m.

4   Convert each time to 12-hour time.
    a   0800   **b**   2230   **c**   0445   **d**   2018
    e   0320   **f**   1058   **g**   2149   **h**   0046

5   Copy and complete this table.

Time	Hours and minutes	Time	Minutes and seconds
5.2 hours		19.5 minutes	
192 minutes		225 seconds	
1246 minutes		386 seconds	
2.25 hours		54.4 seconds	

6   Round:
    a   2h 52 min to the nearest hour   **b**   6 h 23 min to the nearest hour
    c   7.2 hours to the nearest hour   **d**   5.9 hours to the nearest hour
    e   15 min 40s to the nearest minute   **f**   8 min 29 s to the nearest minute
    g   9.8 min to the nearest minute   **h**   22.45 min to the nearest minute

7   Ania boarded a train at 0920 and arrived at her destination at 1345.
    a   How long was her train journey?
    b   If she needed to be home by 1850 the next day and there was a train leaving at 1355
        and at 1458, what is the latest one she could catch?

8   Find the time difference between each pair of times.
    a   2:40 a.m. to 7:45 p.m.   **b**   1:50 a.m. to 11:48 a.m.
    c   10:26 p.m. to 8:32 a.m. the next day **d**   0435 to 1638
    e   1226 to 2045   **f**   1456 to 0912 the next day

9   Darren was baking pastries and placed them in the oven at 5:24 p.m. If they need
    38 minutes to bake, at what time should he take them out?

10 Find the sum of 6 h 28 min, 9 h 54 min and 3.6 hours.

9–05 Timetables

This is a section of train timetable from Central to Mt Victoria on the Blue Mountains line.

Central	7:23j	07:52	08:18j	09:18j	10:18j	11:18j	12:18j	13:18j	13:48j
Redfern	---	07:54	---	---	---	---	---	---	---
Strathfield	7:36	08:06	08:31	09:31	10:31	11:31	12:31	13:31	14:01
Lidcombe	---	08:11	---	---	---	---	---	---	---
Granville	---	08:17	---	---	---	---	---	---	---
Parramatta	7:48	08:21	08:43	09:43	10:43	11:43	12:43	13:43	14:13
Westmead	---	---	---	---	---	---	---	---	---
Blacktown	7:57	08:30	08:52	09:52	10:52	11:52	12:52	13:52	14:22
Penrith	08:11	08:46	09:06	10:06	11:06	12:06	13:06	14:06	14:36
Emu Plains	08:14	08:49	09:09	10:09	11:09	12:09	13:09	14:09	14:39
Lapstone	08:20	08:55	09:15	10:15	11:15	12:15	13:15	14:15	14:45
Glenbrook	08:24	08:59	09:19	10:19	11:19	12:19	13:19	14:19	14:49
Blaxland	08:30	09:05	09:25	10:25	11:25	12:25	13:25	14:25	14:55
Warrimoo	08:33	09:08	09:28	10:28	11:28	12:28	13:28	14:28	14:58
Valley Heights	08:37	09:12	09:32	10:32	11:32	12:32	13:32	14:32	15:02
Springwood	08:42	09:17	09:37	10:37	11:37	12:37	13:37	14:37	15:07
Faulconbridge	08:46	---	09:41	10:41	11:41	12:41	13:41	14:41	---
Linden	08:51	---	09:46	10:46	11:46	12:46	13:46	14:46	---
Woodford	08:55	---	09:50	10:50	11:50	12:50	13:50	14:50	---
Hazelbrook	09:00	---	09:55	10:55	11:55	12:55	13:55	14:55	---
Lawson	09:03	---	09:58	10:58	11:58	12:58	13:58	14:58	---
Bullaburra	09:06	---	10:01	11:01	12:01	13:01	14:01	15:01	---
Wentworth Falls	09:12	---	10:07	11:07	12:07	13:07	14:07	15:07	---
Leura	09:18	---	10:13	11:13	12:13	13:13	14:13	15:13	---
Katoomba	09:22	---	10:17	11:17	12:17	13:17	14:17	15:17	---
Medlow Bath	09:29	---	10:24	11:24	12:24	13:24	14:24	---	---
Blackheath	09:35	---	10:30	11:30	12:30	13:30	14:30	---	---
Mount Victoria	09:42	---	10:37	11:37	12:37	13:37	14:37	---	---

Source: Sydney Trains

EXAMPLE 8

Khodr is meeting Melanie at Springwood at 9:30 a.m. and it is a 7-minute walk from the station to Melanie's house.

a What is the latest train that he could catch from Central to get there on time?

b How long would this train journey take?

c If Khodr caught the next train from Central, what time would he arrive at Melanie's house?

SOLUTION

a Working backwards, if Khodr needs to be at Melanie's house by 9:30 a.m., he needs to arrive at Springwood station by 9:30 a.m. – 7 min = 9:23 a.m.

One train arrives at Springwood at 9:17 a.m., and from the timetable, it departs Central at 7:52 a.m.

So Khodr would need to catch the 7:52 a.m. train from Central.

b The journey lasts from 7:52 a.m. to 9:17 a.m., which is 1 h 25 min.

c The next train that goes from Central to Springwood leaves at 8:18 a.m. and arrives at Springwood at 9:37 a.m. Khodr would arrive at Melanie's house 7 minutes later at 9:44 a.m.

ISBN 9780170351058

Use the timetable on the previous page to answer each question.

1   If Mark needs to be at Wentworth Falls by 11 a.m. and he lives at Blaxland, what is the latest train that he can catch from Blaxland? Select the correct answer **A**, **B**, **C** or **D**.

   A 10:25   **B** 9:25   **C** 10:19   **D** 10:28

2   If Lucy needs to be at Penrith by 9 a.m., what time will the latest train arrive in Penrith if she catches a train from Parramatta? Select **A**, **B**, **C** or **D**.

   A 08:11   **B** 09:06   **C** 08:06   **D** 08:46

3   Find which is the latest train to catch and how long the journey will take if you have to be at:

   a   Valley Heights by 9 a.m. travelling from Central
   b   Penrith by 11:15 a.m. travelling from Strathfield
   c   Glenbrook by 9:15 a.m. travelling from Redfern
   d   Lapstone by 10:30 a.m. travelling from Parramatta
   e   Woodford by 1:55 p.m. travelling from Blacktown
   f   Leura by 3:10 p.m. travelling from Blaxland

4   If Dinesh caught the train from Emu Plains at 1:09 p.m., where would he be:

   a   after 6 min?   **b**   at 1:32 p.m.?   **c**   after 19 min?   **d**   at 1:46 p.m.?

5   Would you be late if you caught the:

   a   9:18 train from Central for a meeting at Blaxland at 10:30 a.m.?
   b   10:43 train from Parramatta for an appointment at Hazelbrook at 11:45?
   c   8:37 train from Valley Heights for a connection at 9:25 a.m. at Katoomba?
   d   14:36 train from Penrith for a meeting at 3 p.m. at Springwood?

6   **a**   How many trains from Central do not stop at Blackheath? At what times do they leave Central?

   b   What would be the 2 major train stations after Central? How do you know?

7   Harry is flying from Brisbane to Singapore and the details of his flight are shown here.

Departure city	Departure time	Boarding gate	Arrival time	Destination
Brisbane	1056	24	1948	Singapore

   a   If Harry has to be at the airport $2\frac{1}{2}$ hours before his flight leaves, what time should he arrive at Brisbane airport? Answer in 12-hour time.
   b   How long will the flight take if there are no delays?
   c   What time will Harry arrive in Singapore if the flight is 25 minutes late? Answer in 12-hour time.

International time zones

The world is divided into 24 different time zones, each one representing a 1 hour time difference. World times are measured in relation to the Greenwich Observatory in London, either ahead or behind UTC (Coordinated Universal Time), also known as GMT (Greenwich Mean Time).

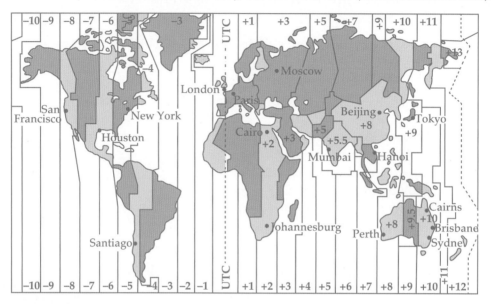

EXAMPLE 9

If it is 10:30 a.m. in Greenwich, then find the time in:

a   Brisbane   b   Mumbai   c   San Francisco

SOLUTION

a   Brisbane is 10 hours ahead of UTC, so its time is 10:30 a.m. + 10 h = 8:30 p.m.

b   Mumbai is 5.5 hours ahead of UTC, so its time is 10:30 a.m. + 5.5 h = 4 p.m.

c   San Francisco is 8 hours behind UTC so its time is 10:30 a.m. − 8 h = 2:30 a.m.

Australia has three time zones: Western (UTC + 8), Central (UTC + 9.5) and Eastern (UTC + 10).

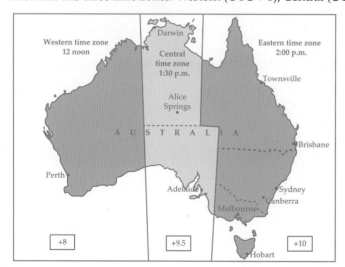

ISBN 9780170351058

EXAMPLE 10

If it is 9:30 a.m. in Alice Springs, find the time in:

a Perth   b Sydney   c Darwin

SOLUTION

a The time in Perth is 1.5 hours behind the time in Alice Springs, so its time is
   9:30 a.m. – 1.5 h = 8 a.m.

b The time in Sydney is half an hour ahead of the time in Alice Springs, so its
   time is 9:30 a.m. + 0.5 h = 10 a.m.

c Darwin is in the same time zone as Alice Springs, so its time is 9:30 a.m.

EXERCISE 9-06

1 If it is 11:30 a.m. in Greenwich, what time will it be in Johannesburg, South Africa? Select
   the correct answer **A, B, C** or **D**.

   A 12:30 p.m.   **B** 9:30 a.m.   **C** 10:30 a.m.   **D** 1:30 p.m.

2 If it is 6:15 pm in Beijing, China, what time will it be in London? Select **A, B, C** or **D**.

   A 2:15 p.m.   **B** 8:15 p.m.   **C** 10:15 a.m.   **D** 2:15 a.m.

3 When it is 8 a.m. in Greenwich, find the time in each city.

   a Tokyo, Japan   b Perth   c Santiago, Chile d Cairns

4 When it is 2:30 p.m. in Sydney, find the time in each city.

   a London   b New York, USA

   c Houston, USA   d Johannesburg, South Africa

5 Laura caught a direct flight from Sydney to Paris. She left Sydney at 9:20 a.m. on Friday
   and the flight was 28 hours long. What was the local time and day when she arrived in
   Paris?

6 Luke watched the Wimbledon tennis final live on TV from his home in Brisbane. The
   match was played in London at 6.00 p.m. and lasted 3 hours 18 minutes. Between what
   times did Luke watch the match in Brisbane?

7 If it is 4:30 p.m. in Adelaide, find the time in each city.

   a Brisbane   b Broome   c Townsville   d Melbourne

8 If it is 7:10 a.m. in Perth, find the time in each city.

   a Sydney   b Alice Springs   c Cairns   d Hobart

9 Each year, most Australian states change to daylight saving time from October to March
   to take advantage of the longer hours of daylight. Clocks are turned forward one hour
   during this period, ahead of standard time. When it is 3 p.m. in the eastern states in
   February, what time is it in Queensland where daylight saving does not operate?

10 Dylan and Kate travelled to Cairns on a flight that left Sydney at 11:30 a.m. daylight saving
   time. The flight was 3.5 hours long. What was the local time in Cairns when they arrived?

CODE PUZZLE

Use this table to decode the answers to the questions below.

A	B	C	D	E	F	G	H	I	J	K	L	M
1	2	3	4	5	6	7	8	9	10	11	12	13

N	O	P	Q	R	S	T	U	V	W	X	Y	Z
14	15	16	17	18	19	20	21	22	23	24	25	26

1   What do you get if you divide the length of a marathon race by the time it takes a marathon runner to finish the race?

20	8	5	/	1	22	5	18	1	7
5	/	19	16	5	5	4	/	15	6
/	20	8	5	/	13	1	18	1	20
8	15	14	/	18	21	14	14	5	18

2   What does a marathon runner have in common with time?

20	8	5	/	12	15	14	7	5	18	/
20	8	5	/	18	1	3	5	/	20	8
5	/	13	15	18	5	/	20	9	13	5
/	8	5	/	20	1	11	5	19		

3   What did the marathon runner say at the end of the race?

9	20	19	/	1	2	15	21	20	/	20
9	13	5	/	9	/	6	9	14	9	19
8	5	4								

PRACTICE TEST 9

Part A General topics

Calculators are not allowed.

1   Evaluate $66\frac{2}{3}$% of $360.

2   Convert $\frac{23}{4}$ to a mixed numeral.

3   Simplify $\dfrac{\left(x^2\right)^4}{x^3}$

4   Find the average of –4, 6, 8 and –2.

5   Find a simplified algebraic expression for the perimeter of this triangle.

6   What type of triangle is shown in Question **5**?

7   Which congruence test proves that $\triangle PSR \equiv \triangle PQR$?

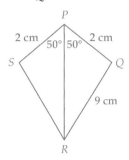

8   Simplify 28 : 16

9   Simplify 9^{-2}

10   Solve $6x + 26 = x + 1$.

Part B Length and time

Calculators are allowed.

9–01 Perimeter

11   What is the perimeter of a rectangle with length 8.9 cm and width 5.6 cm? Select the correct answer **A, B, C** or **D**.

   A   14.5 cm   **B**   19.1 cm   **C**   23.4 cm   **D**   29 cm

12   What is the perimeter of a square with side length 11.7 m? Select **A, B, C** or **D**.

   A   23.4 m   **B**   46.8 m   **C**   35.1 m   **D**   47.2 m

9–02 Parts of a circle

13   Name each part of a circle shown.

   a

   b

   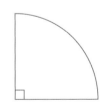

9–03 Circumference of a circle

14 Find, correct to two decimal places, the circumference of each circle.

a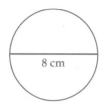

8 cm

b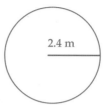

2.4 m

15 Find, correct to two decimal places, the perimeter of this shape.

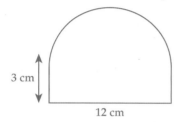

3 cm

12 cm

9–04 Time

16 Find the time difference between:
    a 7:35 p.m. and 1:15 a.m. the next day **b** 0855 and 1924

17 Round to the nearest hour.
    a 11 h 37 min   **b** 7.35 h

9–05 Timetables

Use the timetable on page 178 to answer the questions below.

18 a What is the latest train that I can catch from Central if I have to meet a friend in Emu Plains at 10:15 a.m.?

    b How long would this train journey take?

19 What train should I catch from Strathfield if I had an appointment at Blaxland at 12:30 p.m.?

9–06 International time zones

20 Use the time zones on page 180 to answer the questions below.
    a If it is 2.50 a.m. in New York, what time is it in Sydney?
    b If it is 4:25 p.m. in Townsville, what time is it in Darwin?

INDICES

10

IN THIS CHAPTER YOU WILL:

- multiply and divide terms with the same base
- find a power of a power
- use zero and negative indices
- round numbers to significant figures
- understand and use scientific notation for large and small numbers

Shutterstock.com/ Triff

WORDBANK

base The main number that is raised to a power; for example, in 4^3 the base is 4.

power The number at the top right corner of the base that represents repeated multiplication by itself; for example, 4^3 means $4 \times 4 \times 4$ and the power is 3.

index Another word for power.

indices The plural of index, pronounced 'in-der-sees'.

To multiply terms with the same base, add the indices.
$a^m \times a^n = a^{m+n}$

To divide terms with the same base, subtract the indices.
$a^m \div a^n = \dfrac{a^m}{a^n} = a^{m-n}$

EXAMPLE 1

Simplify each expression, writing the answer in index notation.

a $7^3 \times 7^5$ **b** $x \times x^7$ **c** $5w^2 \times 3w^4$ **d** $7m^6 \times (-4m^3)$

e $5^9 \div 5^3$ **f** $n^{11} \div n^4$ **g** $\dfrac{24w^{12}}{6w^6}$ **h** $-32x^{15} \div 4x^5$

SOLUTION

Add the indices if the bases are the same.

a $7^3 \times 7^5 = 7^8$

*   Add the indices if the bases are the same

b $x \times x^7 = x^8$ ←——— $x = x^1$

c $5w^2 \times 3w^4 = 5 \times 3 \times w^2 \times w^4$
$$= 15w^6$$

d $7m^6 \times (-4m^3) = 7 \times (-4) \times m^6 \times m^3$
$$= -28m^9$$

e $5^9 \div 5^3 = 5^6$

f $n^{11} \div n^4 = n^7$

g $\dfrac{24w^{12}}{6w^6} = \dfrac{24}{6} \dfrac{w^{12}}{w^6}$
$$= 4w^6$$

h $-32x^{15} \div 4x^5 = \dfrac{-32x^{15}}{4x^5}$
$$= -8x^{10}$$

1 Simplify $4^2 \times 4^5$. Select the correct answer **A**, **B**, **C** or **D**.

   A 4^{10} **B** 16^{10} **C** 4^7 **D** 16^7

2 Simplify $6^{18} \div 6^9$. Select **A**, **B**, **C** or **D**.

   A 6^9 **B** 1^9 **C** 6^2 **D** 1^2

3 Copy and complete each statement.

   a 8^3 means _____   **b** The base of 8^3 is _____   **c** The index of 8^3 is _____

   d $2^8 \times 2^3 = 2^{8+\square} = 2^\square$ **e** $x^9 \times x^3 = x^{\square+3} = x^\square$ **f** $12^6 \times 12^3 = 12^{6+\square} = 12^\square$

   g $3^8 \div 3^3 = 3^{8-\square} = 3^\square$ **h** $\dfrac{m^8}{m^2} = m^{\square-2} = m^\square$ **i** $9^6 \div 9^2 = 9^{6-\square} = 9^\square$

4 Simplify each expression.

   a $9^3 \times 9^4$ **b** $a^3 \times a^5$ **c** $n^2 \times n^5$ **d** $w^3 \times w^5$

   e $v^3 \times v^8$ **f** $n^9 \times n^4$ **g** $a^3 \times a^{12}$ **h** $6a^2 \times a^5$

   i $2n^3 \times 8n^4$ **j** $-4a^3 \times 8a^4$ **k** $9n^2 \times (-5n^5)$ **l** $-6w^3 \times (-4w^5)$

   m $-2v^5 \times 5v^4$ **n** $7n^3 \times (-3n^4)$ **o** $-5a^3 \times (-6a^5)$ **p** $12a^2 \times (-4a^5)$

5 Simplify each expression.

   a $9^{11} \div 9^4$ **b** $a^{12} \div a^4$ **c** $\dfrac{n^6}{n^5}$ **d** $w^{15} \div w^5$

   e $v^8 \div v^4$ **f** $n^8 \div n^4$ **g** $a^7 \div a^4$ **h** $\dfrac{6a^{12}}{a^4}$

   i $12n^6 \div 4n^4$ **j** $8a^8 \div 2a^4$ **k** $\dfrac{15n^6}{5n^3}$ **l** $16w^{15} \div (-8w^5)$

   m $-18v^{12} \div 6v^4$ **n** $\dfrac{21n^8}{-3n^4}$ **o** $-63a^7 \div (-9a^4)$ **p** $24a^{12} \div (-4a^5)$

6 Is each equation true or false?

   a $3^6 \times 3^3 = 3^9$ **b** $x^{18} \div x^6 = x^3$

   c $\dfrac{6^8}{6^3} = 6^5$ **d** $n^8 \times 4n^3 = 4n^{11}$

7 **a** Can $8^2 \times 7^3$ and $3^5 \div 2^3$ be simplified using index laws?

   b Evaluate each expression in part **a**.

8 Simplify each expression.

   a $-3bc^2 \times 8b^2c^4$ **b** $4a^3b \times 9a^4b^3$

   c $-3m^3n \times 4m^4n^5$ **d** $-3a^3c \times (-7a^4c^4)$

   e $-24b^8c^6 \div 8b^2c^4$ **f** $\dfrac{45a^{12}b^9}{-9a^4b^3}$

   g $-16m^8n^7 \div 4m^4n^5$ **h** $\dfrac{-21a^9c^6}{-7a^4c^4}$

10-02 Power of a power

To find a power of a power, multiply the indices.

$(a^m)^n = a^{m \times n} = a^{mn}$

EXAMPLE 2

Simplify each expression, giving each answer in index notation.

a $(2^5)^3$ **b** $(x^2)^4$ **c** $(n^6)^0$

SOLUTION

a $(2^5)^3 = 2^{5 \times 3}$ **b** $(x^2)^4 = x^{2 \times 4}$ **c** $(n^6)^0 = n^{6 \times 0}$

$\qquad = 2^{15}$ $\qquad\qquad = x^8$ $\qquad\qquad = n^0$

 Multiply the indices

To find a power of ab or $\dfrac{a}{b}$, raise a and b to the power separately.

$(ab)^n = a^n b^n$ and $\left(\dfrac{a}{b}\right)^n = \dfrac{a^n}{b^n}$

EXAMPLE 3

Simplify each expression.

a $(4x^3)^2$ **b** $\left(\dfrac{x^4}{3}\right)^3$ **c** $\left(\dfrac{5a}{b^6}\right)^2$

SOLUTION

Always raise each part in the bracket to the power separately.

a $(4x^3)^2 = 4^2 \times (x^3)^2$

$\qquad = 16x^6$

b $\left(\dfrac{x^4}{3}\right)^3 = \dfrac{(x^4)^3}{3^3}$

$\qquad = \dfrac{x^{12}}{27}$

c $\left(\dfrac{5a}{b^6}\right)^2 = \dfrac{5^2 a^2}{(b^6)^2}$

$\qquad = \dfrac{25a^2}{b^{12}}$

Shutterstock.com/Rich Carey

1 Simplify $(3^3)^2$. Select the correct answer **A**, **B**, **C** or **D**.
   A 3^5 **B** 27^5 **C** 3^6 **D** 27^2

2 Simplify $(2a^4)^2$. Select **A**, **B**, **C** or **D**.
   A $2a^6$ **B** $4a^8$ **C** $4a^6$ **D** $2a^8$

3 Copy and complete each equation.
   a $(4^3)^2 = 4^{3 \times \underline{\quad}} = 4^{\underline{\quad}}$ **b** $(x^5)^3 = x^{\underline{\quad} \times 3} = \underline{\quad}^{15}$

4 Simplify each expression, giving each answer in index notation.
   a $(2^4)^3$ **b** $(5^2)^4$ **c** $(4^4)^3$ **d** $(3^6)^3$
   e $(9^8)^5$ **f** $(x^3)^2$ **g** $(n^3)^5$ **h** $(m^7)^2$
   i $(w^3)^5$ **j** $(a^6)^4$ **k** $(7^5)^3$ **l** $(x^2)^6$
   m $(5^4)^7$ **n** $(q^6)^5$ **o** $(p^8)^9$

5 Is each statement true or false?
   a $(3a^4)^3 = 3a^{12}$ **b** $(2x^2)^3 = 8x^6$ **c** $(6n^4)^2 = 36n^8$

6 Simplify each expression.
   a $(3x^3)^2$ **b** $(2a^4)^3$ **c** $(4n^6)^2$ **d** $(3m^8)^3$
   e $(5c^5)^2$ **f** $(3w^7)^4$ **g** $(7b^4)^3$ **h** $(8t^5)^2$
   i $(-2a^6)^3$ **j** $(5w^6)^3$ **k** $(-3c^8)^2$ **l** $(-2q^8)^5$

7 Is each statement true or false?
   a $\left(\dfrac{n^2}{2}\right)^3 = \dfrac{n^6}{8}$ **b** $\left(\dfrac{4}{w^6}\right)^2 = \dfrac{4}{w^{12}}$ **c** $\left(\dfrac{3a^3}{2}\right)^4 = \dfrac{3a^{12}}{16}$

8 Simplify each expression.
   a $\left(\dfrac{x^4}{2}\right)^3$ **b** $\left(\dfrac{m^3}{4}\right)^2$ **c** $\left(\dfrac{3}{w^5}\right)^2$ **d** $\left(\dfrac{3}{m^8}\right)^2$
   e $\left(\dfrac{x^9}{3}\right)^3$ **f** $\left(\dfrac{w^5}{-2}\right)^3$ **g** $\left(\dfrac{3}{c^4}\right)^3$ **h** $\left(\dfrac{-2}{n^7}\right)^3$

9 Is each statement true or false?
   a $\left(\dfrac{3a^3}{-2}\right)^2 = \dfrac{9a^6}{-8}$ **b** $\left(\dfrac{a^3}{b^4}\right)^4 = \dfrac{a^{12}}{b^{16}}$

10 Simplify each expression.
   a $\left(\dfrac{2x^6}{3}\right)^3$ **b** $\left(\dfrac{m^5}{n^2}\right)^4$ **c** $\left(\dfrac{2^2}{3a^4}\right)^3$ **d** $\left(\dfrac{a^9}{b^{14}}\right)^4$

What does $3^5 \div 3^5$ equal?

$$3^5 \div 3^5 = \frac{3 \times 3 \times 3 \times 3 \times 3}{3 \times 3 \times 3 \times 3 \times 3} = 1$$

But also, when dividing terms with the same base, we subtract indices:

$3^5 \div 3^5 = 3^{5-5} = 3^0$

So $3^0 = 1$.

What does $3^4 \div 3^6$ equal?

$$3^4 \div 3^6 = \frac{3 \times 3 \times 3 \times 3}{3 \times 3 \times 3 \times 3 \times 3 \times 3} = \frac{1}{3^2}$$

But also, when dividing terms with the same base, we subtract indices:

$3^4 \div 3^6 = 3^{4-6} = 3^{-2}$

So $3^{-2} = \dfrac{1}{3^2}$.

Any term raised to the **power of 0** is 1.

$a^0 = 1$

A term raised to a **negative power** gives a fraction with numerator 1 and denominator the same term raised to a positive power.

$a^{-n} = \dfrac{1}{a^n}$

EXAMPLE 4

Simplify each expression.

a 7^0

b $2a^0$

c $(5n)^0$

d $(-3w)^0$

e 4^{-2}

f x^{-3}

g $3a^{-5}$

h $\dfrac{1}{3}m^{-5}$

SOLUTION

a $7^0 = 1$

b $2a^0 = 2 \times 1$
$ = 2$

c $(5n)^0 = 1$

d $(-3w)^0 = 1$

 ✳ Any number or pronumeral to the power of 0 is 1.

e $4^{-2} = \dfrac{1}{4^2}$
$\phantom{4^{-2}} = \dfrac{1}{16}$

f $x^{-3} = \dfrac{1}{x^3}$

g $3a^{-5} = 3 \times \dfrac{1}{a^5}$
$\phantom{3a^{-5}} = \dfrac{3}{a^5}$

h $\dfrac{1}{3}m^{-5} = \dfrac{1}{3} \times \dfrac{1}{m^5}$
$\phantom{\dfrac{1}{3}m^{-5}} = \dfrac{1}{3m^5}$

 ✳ Any number or pronumeral raised to a negative power gives a fraction:

$a^{-n} = \dfrac{1}{a^n}$

1 Simplify $4a^{-2}$. Select the correct answer **A**, **B**, **C** or **D**.

   A $\dfrac{1}{4a^2}$ **B** $\dfrac{4}{a^2}$ **C** $\dfrac{16}{a^2}$ **D** $\dfrac{1}{16a^2}$

2 Write $\dfrac{2}{x^5}$ with a negative index. Select **A**, **B**, **C** or **D**.

   A $32x^{-5}$ **B** $5x^{-2}$ **C** $2x^{-5}$ **D** $25x^{-5}$

3 Is each statement true or false?

   a $3^0 = 1$ **b** $8^0 = 8$ **c** $3n^0 = 1$ **d** $(3n)^0 = 1$

4 Simplify each expression.

   a 2^0 **b** 9^0 **c** $(-4)^0$

   d $(-10)^0$ **e** n^0 **f** x^0

   g 6^0 **h** $5x^0$ **i** $(2a)^0$

   j $7m^0$ **k** $(-3n)^0$ **l** $7a^0 + 3^0$

   m 3×4^0 **n** $5 \times x^0$ **o** $8a^0 \times m^0$

   p $4w^0 \times 5^0$ **q** $5^0 + 7^0$ **r** $4^0 + 3a^0$

   s $5a^0 + n^0$ **t** $2x^0 + 4y^0$ **u** $6^0 - 2^0$

   v $3x^0 - a^0$ **w** $4a^0 - 3b^0$ **x** $(2a^0)^2$

5 Copy and complete each equation.

   a $5^{-3} = \dfrac{1}{5^{\square}}$ **b** $x^{-4} = \dfrac{\square}{x^{\square}}$ **c** $3n^{-2} = \dfrac{3}{n^{\square}}$

6 Simplify each expression.

   a 2^{-3} **b** 3^{-2} **c** 5^{-4} **d** 4^{-3}

   e n^{-5} **f** x^{-2} **g** m^{-4} **h** $5x^{-1}$

   i a^{-3} **j** $(5a)^{-2}$ **k** $2a^{-4} \times 4a^{-2}$ **l** n^{-8}

   m $(3b)^{-3}$ **n** $10b^{-6} \div 5b^{-2}$ **o** $6x^{-5}$ **p** $(4n)^{-2}$

   q $(3a^4)^{-3}$ **r** $8b^{-3}$ **s** $(2c)^{-5}$ **t** $m^{-2} \times 6m^{-3}$

7 Simplify $(6x^{-2})^0$ and $6(x^{-2})^0$. Are they the same?

8 Simplify each expression.

   a $(7w^0)^{-2}$ **b** $(4a^{-3})^0$

   c $12a^{-4} \div 3a^{-2}$ **d** $4b^{-2} \times 2b^{-3}$

Note that these index laws apply only to terms that have the **same base**.

When multiplying, add indices	$a^m \times a^n = a^{m+n}$
When dividing, subtract indices	$a^m \div a^n = a^{m-n}$
To find a power of a power, multiply indices	$(a^m)^n = a^{m \times n}$
To raise ab to a power:	$(ab)^n = a^n b^n$
To raise $\dfrac{a}{b}$ to a power:	$\left(\dfrac{a}{b}\right)^n = \dfrac{a^n}{b^n}$
To find a zero index:	$a^0 = 1$
To find a negative index:	$a^{-n} = \dfrac{1}{a^n}$

EXAMPLE 5

Simplify each expression.

a $4a^2b^4 \times 3ab^6$

b $27mn^7 \div (-3mn^{-4})$

c $(5b^4)^2 \times (3b)^0$

SOLUTION

a $\begin{aligned}4a^2b^4 \times 3ab^6 &= 4 \times 3 \times a^2a \times b^4b^6 \\ &= 12a^3b^{10}\end{aligned}$

b $\begin{aligned}27mn^7 \div (-3mn^{-4}) &= \frac{27}{-3} \frac{m}{m} \frac{n^7}{n^{-4}} \\ &= -9 \times 1 \times n^{11} \\ &= -9n^{11}\end{aligned}$

c $\begin{aligned}(5b^4)^2 \times (3b)^0 &= 5^2b^8 \times 1 \\ &= 25b^8\end{aligned}$

Corbis Super RF / Alamy

1 When multiplying terms with the same base, the indices are what? Select the correct answer **A**, **B**, **C** or **D**.

    A added   **B** subtracted   **C** multiplied   **D** divided

2 When finding a power of a power, the indices are what? Select **A**, **B**, **C** or **D**.

    A added   **B** subtracted   **C** multiplied   **D** divided

3 Is each statement true or false?

    a $3a^4 \times 5a^5 = 15a^{20}$ **b** $24m^6 \div 8m^3 = 3m^9$

    c $4x^0 = 1$ **d** $(2n^6)^3 = 8n^{18}$

4 Simplify each expression.

    a $b^3 \times b^2$ **b** $n^4 \times n^3$ **c** $a^5 \times a^6$ **d** $w^8 \times w^4$

    e $5x^4 \times 3x^5$ **f** $4a^4 \times 3a^9$ **g** $5m^3 \times 9m^6$ **h** $4v^5 \times 6v^9$

    i $n^9 \div n^3$ **j** $c^5 \div c^4$ **k** $\dfrac{x^8}{x^4}$ **l** $v^{12} \div v^6$

    m $6a^{12} \div 2a^6$ **n** $\dfrac{15m^9}{5m^6}$ **o** $21c^8 \div 7c^2$ **p** $16w^{16} \div 4w^8$

5 Simplify each expression.

    a 3^0 **b** b^0 **c** $4x^0$ **d** $5^0 - 3a^0$

    e $(x^5)^4$ **f** $(a^5)^7$ **g** $(3x^3)^5$ **h** $(2a^6)^3$

    i $(3x^0)^4$ **j** $(6c^7)^0$ **k** $3a^0 \times 4c^0$ **l** $(4w^0)^3 \times 4^0$

6 Simplify each expression.

    a $n^2 \times n^3$ **b** $c^8 \div c^2$ **c** $(x^3)^4$ **d** 7^0

    e $3a^4 \times 5a^6$ **f** $32m^{12} \div 4m^3$ **g** $(2n^5)^3$ **h** $5a^0 + 3x^0$

7 Is each statement true or false?

    a $w^7 \times w^8 = w^{56}$ **b** $3^0 = 3$

    c $v^{16} \div v^8 = v^8$ **d** $8n^{16} \div 4n^4 = 2n^4$

    e $(c^6)^6 = c^{36}$ **f** $2x^5 \times 4x^3 = 8x^{15}$

    g $(2a^0)^4 = 16$ **h** $\dfrac{24c^{12}}{6c^4} = 4c^8$

    i $(4a^6)^3 = 16a^{18}$ **j** $6b^6 \times 8b^9 = 48b^{15}$

8 Simplify each expression.

    a $3a^{-2} \times 4a^{-3}$ **b** $10v^{-4} \div 5v^{-2}$

    c $(7c^{-3})^0$ **d** $(4c^{-2})^{-4}$

    e $3a^5b^7 \times (-6a^0b^{-2})$ **f** $\dfrac{-18m^9n^{-3}}{3m^{-3}n^{-4}}$

    g $7w^0 \times (-3w^5)^3$ **h** $54x^8y^4 \div (-9x^2y^{-2})$

We already know how to round numbers to **decimal places**, but we can also round to **significant figures**.

- The significant figures in the number 13 740 000 are 1, 3, 7 and 4 because they indicate the size of the number. The 0s at the end are not significant.
- The significant figures in the number 9 056 300 are 9, 0, 5, 6 and 3, but the two 0s at the end are not significant.
- The significant figures in the decimal 0.0428 are 4, 2 and 8 because they indicate the size of the decimal. The 0s at the start are not significant.
- So 0s at the end of a whole number or at the start of a decimal are not significant.

EXAMPLE 6

Round each number to 3 significant figures.

a   83 494   b   3 172 000   c   50 437 000

SOLUTION

a   83 494 ≈ 83 500   ⟵——— The 3rd digit 4 is rounded up to 5 as the next digit 9 > 5.

✱ All the 0s at the end of each number are placeholders and are not significant.

b   3 172 000 ≈ 3 170 000   ⟵——— The 3rd digit 7 is rounded down as the next digit 2 < 5.

c   50 437 000 ≈ 50 400 000   ⟵——— The 3rd digit 4 is rounded down as the next digit 3 < 5.

EXAMPLE 7

Write each number correct to 2 significant figures.

a   0.000 629   b   0.005 048 1   c   5.072

SOLUTION

Count the first 2 non-zero digits in each number.

a   0. 000 629 ≈ 0. 000 63   ⟵——— The 2nd significant figure 2 is rounded up to 3 as the next digit 9 > 5.

✱ All the 0s at the start of each number are placeholders and are not significant.

b   0. 005 048 1 ≈ 0. 0050   ⟵——— The 2nd significant figure 0 is rounded down as the next digit 4 < 5.

c   5.072 ≈ 5.1   ⟵——— The 2nd significant figure 0 is rounded up to 1 as the next digit 7 > 5.

Getty Images/Mait Juriado photo

1 Write 32 724 correct to 2 significant figures. Select the correct answer **A, B, C** or **D**.

    A 32 000   **B** 33

    C 32   **D** 33 000

2 Round 0.008 034 0 to 3 significant figures. Select **A, B, C** or **D**.

    A 0.008   **B** 0.008 03

    C 0.008 34   **D** 0.008 034

3 Round each number correct to 3 significant figures.

    a 18 625   b 423 841

    c 75 420   d 1588

    e 3 226 000   f 52 627

    g 48 539 000   h 327 932

    i 207 620   j 17 084 000

    k 90 327 128   l 5 094 000

4 Write each number correct to 2 significant figures.

    a 0.054 18   b 0.002 91

    c 0.7361   d 8.4600

    e 0.002 050 0   f 0.0841

    g 0.000 826   h 0.5048

    i 3.2075   j 0.020 600

    k 29.038   l 0.608 26

5 At the World Cup for soccer in Brazil, there were 94 622 people in the stadium. Round this number correct to:

    a 1 significant figure

    b 2 significant figures

    c 3 significant figures

6 The concentration of medicine in a cup of water was 0.0876. Round this number correct to 2 significant figures.

7 The population of Gully Plains was 28 629 last year.

    a If the population increased by 6.4% this year, find the population this year.

    b Write this year's population correct to 4 significant figures.

8 Evaluate $\dfrac{78.85 - 11.964}{1.48 + 12.622}$ correct to 2 significant figures.

9 Decrease $2 578 000 by 3.2%, writing the answer correct to 2 significant figures.

Scientific notation is a way of writing very large or small numbers, using a decimal between 1 and 10 multiplied by a power of 10, for example, $2.86 \times 10^7 = 28\ 600\ 000$.

A number written in scientific notation has the form $m \times 10^n$ where m is a number between 1 and 10 and n is an integer.

To write a large number in scientific notation:
- use its significant figures to write a decimal between 1 and 10
- for the power of 10, count the number of places in the number after the first digit

EXAMPLE 8

Write each number in scientific notation.

a 439 700   **b** 52 300 000   **c** 780 360 000

SOLUTION

a 4<u>39 700</u> = 4.397×10^5 ⟵ 5 places underlined after the first significant figure, 4.

b 5<u>2 300 000</u> = 5.23×10^7 ⟵ 7 places underlined after the first significant figure, 5.

c 7<u>80 360 000</u> = 7.8036×10^8 ⟵ 8 places underlined after the first significant figure, 7.

EXAMPLE 9

Write each number in decimal form.

a 5.8×10^3 **b** 7.2804×10^6 **c** 9×10^{10}

SOLUTION

a $5.8 \times 10^3 = 5\ 800$ ⟵ Move the decimal point 3 places right or make 3 places after the 5.

b $7.2804 \times 10^6 = 7\ 280\ 400$ ⟵ Move the decimal point 6 places right or make 6 places after the 7.

c $9 \times 10^{10} = 90\ 000\ 000\ 000$ ⟵ Move the decimal point 10 places right.

EXAMPLE 10

Write these numbers in descending order.

2.6×10^5 2×10^7 2.06×10^5

SOLUTION

To compare numbers in scientific notation, first compare the powers of 10.

Both 2.6×10^5 and 2.06×10^5 are definitely smaller than 2×10^7 because $10^5 < 10^7$.

Since 2.6×10^5 and 2.06×10^5 have the same power of 10, we need to compare their decimal parts.

2.6×10^5 is larger because $2.6 > 2.06$.

So in descending order, the numbers are 2×10^7, 2.6×10^5, 2.06×10^5.

OR: Convert the numbers to decimal form to compare them.

$2.6 \times 10^5 = 260\ 000$ $2 \times 10^7 = 20\ 000\ 000$ $2.06 \times 10^5 = 206\ 000$

From largest to smallest: 20 000 000, 260 000, 206 000.

So in descending order, the numbers are 2×10^7, 2.6×10^5, 2.06×10^5.

1   Write 1 245 000 in scientific notation. Select the correct answer **A, B, C** or **D**.

    A 1.245×10^6 **B** 12.45×10^5 **C** 12.45×10^6 **D** 1.245×10^5

2   Write 7.14×10^4 in decimal form. Select **A, B, C** or **D**.

    A 71 400   **B** 714 000   **C** 7140   **D** 714

3   Copy and complete each equation.

    a $24\,000 = 2.4 \times 10^\square$ **b** $800\,000 = 8 \times 10^\square$

    c $56\,000 = \square \times 10^4$ **d** $756\,000 = \square \times 10^5$

    e $503\,000\,000 = \square \times 10^\square$ **f** $420\,060\,000 = \square \times 10^\square$

4   Write each number in scientific notation.

    a 450 000   **b** 72 600 000 000

    c 568 400 000   **d** 60 800 000

    e 298 000 000   **f** 3.4 million

    g 45 600 000 000   **h** 9.85 billion

    i 6.35 million

5   Copy and complete each equation.

    a $5.4 \times 10^4 = 54...$ **b** $8.16 \times 10^5 = 816...$

    c $9.8 \times 10^8 = 98....$ **d** $2.04 \times 10^2 = 2....$

    e $7.23 \times 10^6 = 723...$ **f** $4.618 \times 10^3 = 46...$

6   Write each number in decimal form.

    a 3.6×10^4 **b** 6.95×10^6

    c 8×10^5 **d** 3.68×10^7

    e 1.276×10^9 **f** 4.2×10^8

    g 1.12×10^{12} **h** 5.8×10^6

    i 6×10^{11}

7   List in ascending order: $6.8 \times 10^6, 8 \times 10^4, 4.002 \times 10^7, 7.48 \times 10^7$

8   List in descending order: $2.6 \times 10^9, 3 \times 10^5, 2.58 \times 10^9, 3.42 \times 10^6$

9   Write each number in scientific notation.

    a There were 93 824 shoppers in the city.

    b There were 258 000 Australians visiting the Olympic Games in 2016.

    c Lake Eucumbene in the Snowy Mountains has a capacity of 6 735 000 000 L.

    d The distance from the Earth to the Moon is about 12 500 000 km.

10 Use the Internet to discover how many bytes are in a gigabyte and write your answer in scientific notation.

A number written in scientific notation has the form $m \times 10^n$ where m is a number between 1 and 10 and n is an integer.

To write a decimal in scientific notation:

- use its significant figures to write a decimal between 1 and 10
- for the **negative** power of 10, count the number of places in the number up to and including the first significant digit (or count the number of 0s)

EXAMPLE 11

Write each number in scientific notation.

a 0.001 68   **b** 0.000 000 8   **c** 0.000 004 2

SOLUTION

a $0.\underline{001}\,68 = 1.68 \times 10^{-3}$ ←—— 3 places underlined up to the first significant figure, 1 (or three 0s)

b $0.\underline{000\,000\,8} = 8 \times 10^{-7}$ ←—— 7 places underlined up to the first significant figure, 8 (or seven 0s).

c $0.\underline{000\,004}\,2 = 4.2 \times 10^{-6}$ ←—— 6 places underlined up to the first significant figure, 4 (or six 0s).

EXAMPLE 12

Write each number in decimal form.

a 6.1×10^{-4} **b** 3.09×10^{-7} **c** 7×10^{-5}

SOLUTION

a $6.1 \times 10^{-4} = 0.000\,61$ ←—— Move the decimal point 4 places left or insert four 0s in front.

✳ The decimal point goes after the first zero.

b 3.09×10^{-7} 0.000 000 309   ←—— Move the decimal point 7 places left or insert seven 0s in front.

c $7 \times 10^{-5} = 0.000\,07$ ←—— Move the decimal point 5 places left or insert five 0s in front.

EXAMPLE 13

List these numbers in ascending order: $8.2 \times 10^{-4}, 8 \times 10^{-3}, 8.04 \times 10^{-4}$

SOLUTION

First compare the powers of 10.

Both 8.2×10^{-4} and 8.04×10^{-4} are smaller than 8×10^{-3} because $10^{-4} < 10^{-3}$.

Since 8.2×10^{-4} and 8.04×10^{-4} have the same power of 10, we need to compare their decimal parts.

8.04×10^{-4} is smaller because $8.04 < 8.2$.

So in ascending order, the numbers are $8.04 \times 10^{-4}, 8.2 \times 10^{-4}, 8 \times 10^{-3}$.

OR: Convert the numbers to decimal form to compare them.

$8.2 \times 10^{-4} = 0.000\,82$ $8 \times 10^{-3} = 0.008$ $8.04 \times 10^{-4} = 0.000\,804$
         $= 0.000\,820$ $= 0.008\,000$ $= 0.000\,804$

From smallest to largest: 0.000 804, 0.000 820, 0.008 000.

So in ascending order, the numbers are $8.04 \times 10^{-4}, 8.2 \times 10^{-4}, 8 \times 10^{-3}$.

ISBN 9780170351058

1   Write 0.00718 in scientific notation. Select the correct answer **A, B, C** or **D**.
    A 7.18×10^{-2} **B** 71.8×10^{-4} **C** 7.18×10^{-3} **D** 718×10^{-5}

2   Write 6.4×10^{-4} in decimal form. Select **A, B, C** or **D**.
    A 0.000 64   **B** 0.064   **C** 0.000 064   **D** 0.0064

3   Copy and complete each equation.
    a $0.002 = 2 \times 10^{\square}$ **b** $0.0032 = 3.2 \times 10^{\square}$
    c $0.000\ 785 = \square \times 10^{-4}$ **d** $0.000\ 004 = \square \times 10^{-6}$
    e $0.000\ 015 = \square \times 10^{\square}$ **f** $0.000\ 003\ 18 = \square \times 10^{\square}$

4   Write each number in scientific notation.
    a 0.0005   **b** 0.000 62   **c** 0.000 070 8
    d 0.007 004   **e** 0.000 000 4   **f** 0.000 000 92
    g 0.000 000 001 86   **h** 0.000 008 6   **i** 0.0023

5   Copy and complete each equation.
    a $5 \times 10^{-3} = 0.\underline{\qquad}$ **b** $7.35 \times 10^{-4} = 0.\underline{\qquad}$
    c $1.9 \times 10^{-5} = 0.\underline{\qquad}$ **d** $2.8 \times 10^{-7} = 0.\underline{\qquad}$
    e $8.04 \times 10^{-1} = 0.\underline{\qquad}$ **f** $4.225 \times 10^{-5} = 0.\underline{\qquad}$

6   Write each number in decimal form.
    a 2.1×10^{-2} **b** 6.28×10^{-5} **c** 5×10^{-4}
    d 4.05×10^{-6} **e** 1.115×10^{-9} **f** 2.6×10^{-7}
    g 3.9×10^{-12} **h** 6.12×10^{-6} **i** 7×10^{-8}

7   List these numbers in ascending order: $3.6 \times 10^{-7}, 4 \times 10^{-3}, 3.06 \times 10^{-7}$

8   List these numbers in descending order: $7.2 \times 10^{-5}, 7 \times 10^{-4}, 7.02 \times 10^{-5}$

9   Write each number in scientific notation.
    a The blink of an eye takes 0.2 s.
    b The length of a fly is 0.0031 m.
    c The amount of poison in a substance is 0.000 007 4 grams.
    d The scale factor of a map is 0.000 025.

10 Use the Internet to discover what decimal a micrometre is of a metre. Write your answer in scientific notation.

To enter a number in scientific notation on a calculator, use the ×10ˣ or EXP key.

EXAMPLE 14

Use a calculator to write each number in decimal form.

a 3.76×10^9

b 2.604×10^{-3}

SOLUTION

a $3.76 \times 10^9 = 3\,760\,000\,000$ ⟵—— On a calculator, enter: 3.76 ×10ˣ 9 =

b $2.604 \times 10^{-3} = 0.002\,604$ ⟵—— On a calculator, enter: 2.604 ×10ˣ (−) 3 =

EXAMPLE 15

Evaluate each expression in scientific notation, correct to two significant figures.

a $(4.2 \times 10^5) \times (7.15 \times 10^8)$

b $\dfrac{8.26 \times 10^8}{1.8 \times 10^{-7}}$

SOLUTION

a On a calculator, enter: 4.2 ×10ˣ 5 × 7.15 ×10ˣ 8 =

$$(4.2 \times 10^5) \times (7.15 \times 10^8) = 3.003 \times 10^{14}$$
$$\approx 3.0 \times 10^{14}$$

b On a calculator, enter: 8.26 ×10ˣ 8 ÷ 1.8 ×10ˣ (−) 7 =

$$\frac{8.26 \times 10^8}{1.8 \times 10^{-7}} = 4.5888.... \times 10^{15}$$
$$\approx 4.6 \times 10^{15}$$

iStockphoto.com/Baoshan Zhang

1 Use a calculator to write 3.15×10^6 in decimal form. Select the correct answer **A**, **B**, **C** or **D**.

  A 315 000 **B** 3 150 000

  C 31 500 **D** 31 500 000

2 Evaluate $(2.6 \times 10^{12}) \times (1.4 \times 10^{-3})$ in decimal form. Select **A**, **B**, **C** or **D**.

  A 364 000 000 **B** 36 400 000 000

  C 3 640 000 000 **D** 36 400 000

3 Use a calculator to write each number in decimal form.

  a 2.9×10^4 **b** 6.28×10^6 **c** 7.1×10^5

  d 4.25×10^{-5} **e** 3.276×10^{-8} **f** 8.6×10^{-4}

  g 9.12×10^{12} **h** 5.08×10^{-7} **i** 5.22×10^{-9}

4 Evaluate each expression in scientific notation, correct to 3 significant figures where necessary.

  a $2.4 \times 10^5 \times 6.2 \times 10^7$ **b** $8 \times 10^6 \times 7.8 \times 10^8$

  c $6.52 \times 10^5 + (8.2 \times 10^2)$ **d** $\dfrac{9.5 \times 10^8}{1.28 \times 10^3}$

  e $7.6 \times 10^8 - (7 \times 10^4)$ **f** $5.14 \times 10^{-3} \times 8.2 \times 10^{-3}$

  g $9.4 \times 10^5 + (3.1 \times 10^6)$ **h** $\dfrac{6 \times 10^5}{2.802 \times 10^{-2}}$

5 Evaluate each expression, correct to 2 significant figures.

  a $\dfrac{7.2 \times 10^8}{1.6 \times 10^4}$ **b** $\dfrac{5.24 \times 10^{12}}{3.6 \times 10^3}$

  c $\dfrac{8.16 \times 10^{12}}{4 \times 10^{-3}}$ **d** $\dfrac{4.06 \times 10^{18}}{3 \times 10^{-4}}$

  e $\dfrac{6.8 \times 10^{16}}{5 \times 10^{-4}}$ **f** $\dfrac{7 \times 10^{-8}}{1.6 \times 10^4}$

  g $\dfrac{9.24 \times 10^{12}}{3.7 \times 10^{-2}}$ **h** $\dfrac{6.2 \times 10^{-12}}{4 \times 10^3}$

6 Evaluate each expression, to 3 significant figures.

  a $(3.65 \times 10^5)^3$ **b** $(4.82 \times 10^6)^4$

  c $\sqrt{8.2 \times 10^{12}}$ **d** $\sqrt{9.16 \times 10^6}$

7 Last year, Zac worked 6 days a week for 48 weeks. Each day, he rode his bicycle 5 km to work and back. Calculate and write down in scientific notation the total distance that Zac covered last year to 3 significant figures.

8 Evaluate $\dfrac{4.5 \times 10^5 \times 7.5 \times 10^{-2}}{9.3 \times 10^8}$ correct to 4 significant figures.

FIND-A-WORD PUZZLE

Make a copy of this puzzle, then find all the words listed below in this grid of letters.

C	O	E	B	L	Y	O	N	S	T	I	Q	X	X	Z
W	A	G	D	X	A	O	K	C	C	V	E	E	V	E
T	N	L	W	I	I	W	S	I	A	J	Z	D	Y	R
G	Q	W	C	T	V	H	S	E	R	O	U	N	D	O
F	M	P	A	U	R	I	D	N	T	I	I	I	M	J
X	R	T	B	E	L	E	D	T	B	Y	E	D	X	F
J	O	A	W	W	C	A	B	I	U	Z	F	J	Y	G
N	S	O	C	I	Y	A	T	F	S	U	I	C	T	M
U	P	D	M	T	S	L	M	I	U	R	G	R	V	H
T	W	A	F	E	I	O	P	C	O	Z	U	F	E	S
P	L	L	T	M	Z	O	N	I	B	N	R	K	B	C
S	E	C	I	D	N	I	N	A	T	K	E	O	H	P
R	O	T	A	L	U	C	L	A	C	L	S	O	V	K
S	I	G	N	I	F	I	C	A	N	T	U	X	D	T
N	E	G	A	T	I	V	E	D	D	A	A	M	B	T

ADD	BASE	CALCULATION	CALCULATOR
DECIMAL	DIVIDE	FIGURES	FRACTION
INDEX	INDICES	LAWS	MULTIPLY
NEGATIVE	NOTATION	POWER	ROUND
SCIENTIFIC	SIGNIFICANT	SUBTRACT	ZERO

ISBN 9780170351058

Part A General topics

Calculators are not allowed.

1 Convert 8.5% to a decimal.

2 Factorise $24ab - 6a^2$

3 What type of triangle is this?

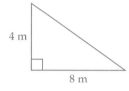
4 m
8 m

4 Find the area of the triangle in Question 3.

5 How many hours are there between 9 a.m. and 5 p.m.?

6 Copy this diagram and mark a pair of co-interior angles.

7 Convert 100 minutes to hours and minutes.

8 Write 1950 in 12-hour time.

9 Write the formula for the area of a circle with radius r.

r

10 If a letter from the word MATHEMATICS is chosen at random, what is the probability that it is a vowel?

Part B Indices

Calculators are allowed.

10–01 Multiplying and dividing terms with the same base

11 Simplify $6n^3 \times 4n^4$. Select the correct answer **A**, **B**, **C** or **D**.

   A $24n^{12}$ **B** $10n^7$ **C** $24n^7$ **D** $10n^{12}$

12 Simplify $32x^8 \div 8x^4$. Select **A**, **B**, **C** or **D**.

   A $24x^4$ **B** $4x^4$ **C** $24x^2$ **D** $4x^2$

10–02 Power of a power

13 Simplify each expression.

   a $(3^5)^4$ **b** $(n^3)^7$ **c** $(2x^3)^5$

14 Simplify each expression.

   a $\left(\dfrac{m^2}{n}\right)^4$ **b** $\left(\dfrac{a^2}{3}\right)^4$ **c** $\left(\dfrac{2^2}{a}\right)^6$

10–03 Zero and negative indices

15 Simplify each expression.

   a 7^0 **b** x^0 **c** $4x^0$

16 Simplify each expression.

   a 5^{-2} b $2a^{-3}$ c $\dfrac{1}{3}x^{-5}$

10–04 Index Laws review

17 Simplify each expression.

   a $3a^2b \times \left(-4ab^2\right)$ b $\dfrac{24m^8n^4}{-8m^5n}$ c $(2a^4)^2 \times 6a^0$

10–05 Significant figures

18 Round each number to 2 significant figures.

   a 516 670   b 4 089 214   c 0.003 277 6

10–06 Scientific notation for large numbers

19 Write each number in scientific notation.

   a 452 200   b 5 920 000   c 23 000 000 000

10–07 Scientific notation for small numbers

20 Write each number in scientific notation.

   a 0.0145   b 0.000 724   c 0.000 000 012

21 List these numbers in ascending order: 7.2×10^6, 6×10^{-4}, 5.6×10^{-4}

10–08 Scientific notation on a calculator

22 Evaluate each expression.

   a $8.35 \times 10^8 \times 1.08 \times 10^{-3}$

   b $\dfrac{12.65 \times 10^{14}}{8.4 \times 10^{-6}}$ correct to 2 significant figures

AREA AND VOLUME

11

IN THIS CHAPTER YOU WILL:

- calculate the areas of triangles, quadrilaterals and composite shapes
- calculate the areas of circles, semicircles, quadrants, sectors, annuluses and composite circular shapes
- calculate the surface areas of cubes, rectangular prisms and triangular prisms
- calculate the surface area of a cylinder
- convert between metric units for volume and capacity
- calculate the volumes of prisms, composite prisms and cylinders

Shutterstock.com/Michael Mihin

Area is the amount of surface space occupied by a flat shape. Area is measured in square units such as square centimetres (cm²) or square metres (m²).

Metric units of area

square centimetre (cm²)	square metre (m²)

Actual size $\boxed{1 \text{ cm}^2}$ 1 cm
 1 cm
 	
 1 cm = 10 mm
 	
 1 cm² = 1 cm × 1 cm
      = 10 mm × 10 mm
      = 100 mm²
	About the size of a large shower recess $\boxed{1 \text{ m}^2}$ 1 m
	 1 m
	
	 1 m = 100 cm
	
	 1 m² = 1 m × 1 m
	      = 100 cm × 100 cm
	      = 10 000 cm²

hectare (ha)	**square kilometre (km²)**

About the size of 2 football fields 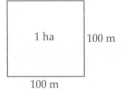 1 ha   100 m
 100 m
 	
 1 ha = 100 m × 100 m
      = 10 000 m²
	About the size of a theme park 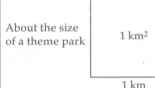 1 km² 1 km
	 1 km
	
	 1 km = 1000 m
	
	 1 km² = 1 km × 1 km
	      = 1000 m × 1000 m
	      = 1 000 000 m² or 100 ha

1 cm² = 100 mm²   1 m² = 10 000 cm²
1 ha = 10 000 m²   1 km² = 1 000 000 m² or 100 ha

EXAMPLE 1

Convert:

a 8 cm² to mm²

b 10.4 ha to m²

c 600 000 m² to km²

d 52 000 cm² to m²

SOLUTION

a 8 cm² = 8 × 100 mm²   ←———— 1 cm² = 100 mm²

         = 800 mm²

b 10.4 ha = 10.4 × 10 000 m²   ←———— 1 ha = 10 000 m²

            = 104 000 m²

c 600 000 m² = (600 000 ÷ 1 000 000) km² ←———— 1 km² = 1 000 000 m²

                  = 0.6 km²

d 52 000 cm² = (52 000 ÷ 10 000) m² ←———— 1 m² = 10 000 cm²

                 = 5.2 m²

Square	Rectangle
$A = (\text{side length})^2$ $A = s^2$	 $A = \text{length} \times \text{width}$ $A = lw$
Triangle 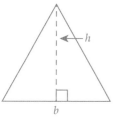 $A = \dfrac{1}{2} \times \text{base} \times \text{height}$ $A = \dfrac{1}{2}bh$	**Parallelogram** 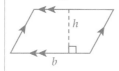 $A = \text{base} \times \text{height}$ $A = bh$

EXAMPLE 2

Find the area of each shape.

a

b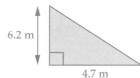

c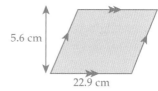

SOLUTION

a $A = lw$ rectangle

  $= 14.6 \times 9.8$

  $= 143.08 \text{ cm}^2$

b $A = \dfrac{1}{2}bh$ triangle

  $= \dfrac{1}{2} \times 4.7 \times 6.2$

  $= 14.57 \text{ m}^2$

c $A = bh$  parallelogram

  $= 22.9 \times 5.6$

  $= 128.24 \text{ cm}^2$

EXAMPLE 3

Find the area of this composite shape.

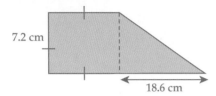

7.2 cm

18.6 cm

SOLUTION

$$\text{Area} = 7.2^2 + \frac{1}{2} \times 18.6 \times 7.2 \quad \longleftarrow \quad \text{Area of square + area of triangle}$$

$$= 118.8 \text{ cm}^2$$

EXERCISE 11-01

1 Find the area of a rectangle with length 16 cm and width 9 cm. Select the correct answer **A**, **B**, **C** or **D**.

    A 50 cm²   **B** 72 cm²   **C** 144 cm²   **D** 288 cm²

2 Find the area of a triangle with base 17 m and height 8 m. Select **A**, **B**, **C** or **D**.

    A 136 m²   **B** 68 m²   **C** 272 m²   **D** 25 m²

3 Copy and complete each equation.

    a 6 m² = _____ cm²   **b** 17 ha = _____ m²

    c 7.4 cm² = _____ mm²   **d** 2.9 km² = _____ m²

    e 4.1 m² = _____ mm²   **f** 10.8 km² = _____ ha

    g 4000 mm² = _____ cm²   **h** 100 000 m² = _____ km²

    i 360 000 m² = _____ ha   **j** 2 000 000 m² = _____ km²

    k 9900 cm² = _____ m²   **l** 580 ha = _____ km²

4 Find the area of each shape.

a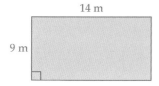
14 m
9 m

b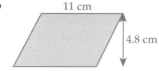
11 cm
4.8 cm

c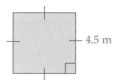
4.5 m

d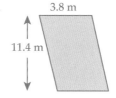
3.8 m
11.4 m

e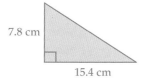
7.8 cm
15.4 cm

f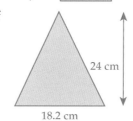
24 cm
18.2 cm

g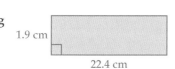
1.9 cm
22.4 cm

h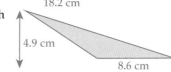
4.9 cm
8.6 cm

5 Find the area of each composite shape.

a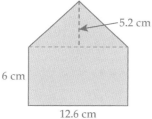
5.2 cm
6 cm
12.6 cm

b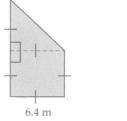
6.4 m

c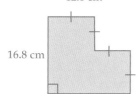
16.8 cm

d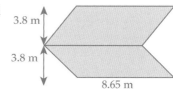
3.8 m
3.8 m
8.65 m

6 a Find the area of a triangular garden with a base of 11.5 m and a height of 8.4 m.

b How many roses can be planted in this garden if each rose bush needs 0.5 m² to grow?

11-02 | Area of a circle

To find the **area of a circle**, multiply π by the circle's radius squared.

$A = \pi r^2$ means $A = \pi \times \text{radius}^2$

If you are given the diameter, halve it to find the radius first.

EXAMPLE 4

Find the area of each circle correct to 2 decimal places.

a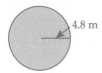
4.8 m

b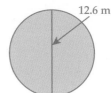
12.6 m

SOLUTION

a $A = \pi r^2$ ⟵ $r = 4.8$

$= \pi \times 4.8^2$

$= 72.38229...$

$\approx 72.38 \text{ m}^2$

b $A = \pi r^2$ ⟵ $r = \frac{1}{2} \times 12.6 = 6.3$

$= \pi \times 6.3^2$

$= 124.6898...$

$\approx 124.69 \text{ cm}^2$

EXAMPLE 5

Find the area of each shape, correct to 1 decimal place.

a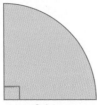
3.4 cm

b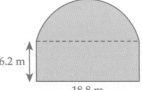
6.2 m
18.8 m

c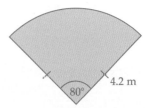
4.2 m
80°

SOLUTION

a Area $= \frac{1}{4} \times \pi \times 3.4^2$

✱ quadrant

$= 9.0792....$

$\approx 9.1 \text{ cm}^2$

b Area $= \frac{1}{2} \times \pi \times 9.4^2 + 18.8 \times 6.2$ $\qquad r = \frac{1}{2} \times 18.8 = 9.4$

✱ semicircle + rectangle

$= 255.3555...$

$\approx 255.4 \text{ m}^2$

c Area $= \frac{80}{360} \times \pi \times 4.2^2$ ⟵ ✱ There are 360° in a circle, but a sector is a fraction of a circle.

$= 12.3150...$

$\approx 12.3 \text{ m}^2$

1 Find the area of a circle with radius 3 cm. Select the correct answer **A, B, C** or **D**.

   A 18.8 cm²   **B** 28.3 cm²   **C** 56.5 cm²   **D** 37.7 cm²

2 Find the area of a circle with diameter 6.8 m. Select **A, B, C** or **D**.

   A 36.3 m²   **B** 72.6 m²   **C** 145.3 m²   **D** 42.7 m²

3 Find the area of each circle correct to 1 decimal place.

   a **b** 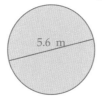 **c**

4 A circular lawn has a diameter of 24.6 m. What is its area to the nearest square metre?

5 Find the area of each shape, correct to 2 decimal places.

   a 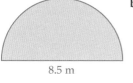 **b** 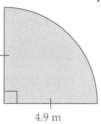 **c**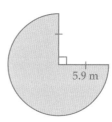

   d 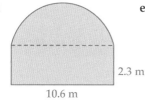 **e** 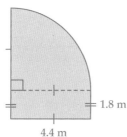 **f**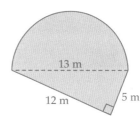

   g 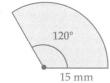 **h** 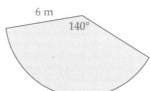 **i**

6 A circular garden has diameter of 3.4 m. Find:

   a the area of the garden, correct to 3 decimal places

   b the number of flowers that can be planted in the garden if they need 0.2 m² each to survive

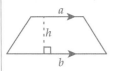

Trapezium	Kite and rhombus

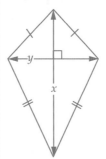

$A = \dfrac{1}{2} \times$ height $\times$ (sum of parallel sides)

$A = \dfrac{1}{2} h(a + b)$

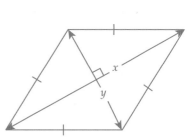

$A = \dfrac{1}{2} \times$ diagonal 1 $\times$ diagonal 2

$A = \dfrac{1}{2} xy$

Shutterstock/rootstock

Dutourdumonde / Alamy

EXAMPLE 6

Find the area of each quadrilateral.

a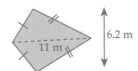

b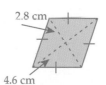

c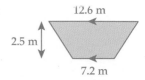

SOLUTION

a $A = \dfrac{1}{2} xy$ ✶ kite

$= \dfrac{1}{2} \times 11 \times 6.2$

$= 34.1 \text{ m}^2$

b $A = \dfrac{1}{2} xy$ ✶ rhombus

$= \dfrac{1}{2} \times 4.6 \times 2.8$

$= 6.44 \text{ cm}^2$

c $A = \dfrac{1}{2} h(a + b)$ ✶ trapezium

$= \dfrac{1}{2} \times 2.5 \times (12.6 + 7.2)$

$= 24.75 \text{ m}^2$

1 Find the area of a kite with diagonals 6 cm and 12.8 cm. Select the correct answer **A**, **B**, **C** or **D**.

   A 76.8 cm²   **B** 153.6 cm²   **C** 38.4 cm²   **D** 37.8 cm²

2 Find the area of a trapezium with parallel sides 5 m and 8.5 m and perpendicular height 9 m. Select **A**, **B**, **C** or **D**.

   A 60.75 m²   **B** 121.5 m²   **C** 43.75 m²   **D** 59.5 m²

3 Find the area of each quadrilateral.

a

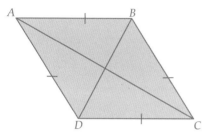

   $AC = 26$ cm
   $BD = 17$ cm

b

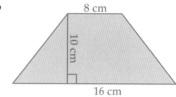

c

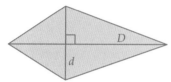

   $D = 20$ cm
   $d = 12$ cm

d

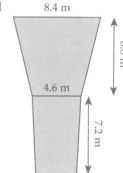

e

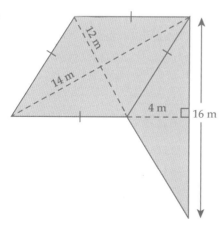

f

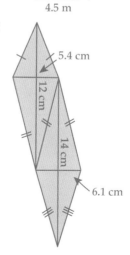

4   Pooja makes a kite using a frame made up of two sticks crossing at right angles. The sticks are 2.6 m and 1.8 m long. The kite is made from canvas costing $36.80 per m². What is the cost of the canvas used to make this kite?

5   A garden is made in the shape of a trapezium with parallel sides 2.4 m and 1.7 m, while the distance between them is 3 m.

   a   What is the area of the garden bed?

   b   How many ferns can be planted in the garden if they need 0.3 m² each?

WORDBANK

surface area The sum of the areas of the faces of a solid shape.

net The faces of a solid shape laid out flat.

The surface area of a solid is easier to calculate by looking at its net. The number of faces and their shape can then be seen.

A cube and its net

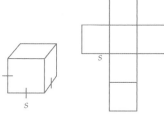

6 identical square faces

A rectangular prism and its net

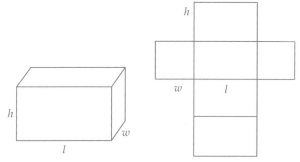

6 rectangular faces, with opposite faces identical: top and bottom, front and back, left and right

---

Surface area of a cube
$SA = 6 \times$ (side length)^2
$A = 6s^2$

Surface area of a rectangular prism
$SA = (2 \times \text{length} \times \text{width}) + (2 \times \text{length} \times \text{height}) + (2 \times \text{width} \times \text{height})$
$SA = 2lw + 2lh + 2wh$

Surface area is an area so it is measured in **square units**.

EXAMPLE 7

Find the surface area of each prism.

a

4 cm

b

8 m
3 m
12 m

SOLUTION

a $SA = 6 \times$ area of squares

$= 6 \times 4^2$

$= 96 \text{ cm}^2$

b $SA = 2 \times \text{bottom} + 2 \times \text{front} + 2 \times \text{sides}$

$= (2 \times 12 \times 3) + (2 \times 12 \times 8) + (2 \times 8 \times 3)$

$= 312 \text{ m}^2$

1   Find the surface area of a cube with side length 7 cm. Select the correct answer **A, B, C** or **D**.

   A 49 cm²   **B** 196 cm²   **C** 294 cm²   **D** 343 cm²

2   What is the surface area of a rectangular prism with length 9 m, width 4 m and height 8 m? Select **A, B, C** or **D**.

   A 288 m²   **B** 140 m²   **C** 208 m²   **D** 280 m²

3   Draw the net of each prism, then find its area.

   a
   5 m

   b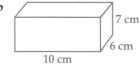
   7 cm
   6 cm
   10 cm

4   Find the surface area of each prism.

   a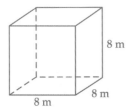
   8 m
   8 m
   8 m

   b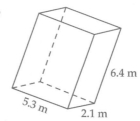
   6.4 m
   5.3 m
   2.1 m

   c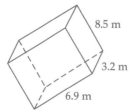
   8.5 m
   3.2 m
   6.9 m

   d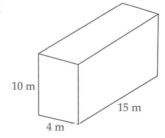
   10 m
   15 m
   4 m

5   **a** Find the surface area of a cube with side length 4.2 mm.

   b Do you think that a rectangular prism with double the length and half the height will have the same surface area as this cube?

   c Check your guess by finding the surface area of the rectangular prism.

A triangular prism and its net

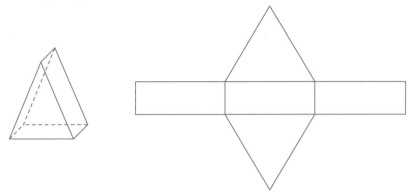

EXAMPLE 8

Find the surface area of the triangular prism with this net.

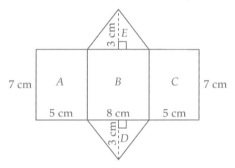

SOLUTION

Faces A and C are identical rectangles, B is another rectangle, D and E are identical triangles.

Surface area $= (2 \times 5 \times 7) + (8 \times 7) + (2 \times \dfrac{1}{2} \times 8 \times 3)$ ⟵ $2 \times$ area A + area B + $2 \times$ area E

$= 150 \text{ cm}^2$

EXAMPLE 9

Find the surface area of the triangular prism.

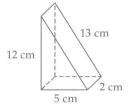

13 cm

12 cm

2 cm

5 cm

SOLUTION

5 faces: front and back are identical triangles, the other 3 faces are different rectangles.

Surface area = Area of 2 triangles + area of base + area of LHS + area of RHS

$= \left(2 \times \dfrac{1}{2} \times 5 \times 12\right) + (5 \times 2) + (12 \times 2) + (13 \times 2)$ ⟵ $2 \times$ front + bottom + sides

$= 120 \text{ cm}^2$

1 What are the shapes of the faces of a triangular prism? Select the correct answer **A, B, C** or **D**.

   A 3 triangles, 2 rectangles   **B** 1 triangle, 4 rectangles

   C 2 triangles, 2 rectangles   **D** 2 triangles, 3 rectangles

2 Find the surface area of a triangular prism if the area of each triangular face is 9 m² and the area of each rectangular face is 15 m². Select **A, B, C** or **D**.

   A 57 m²   **B** 63 m²   **C** 48 m²   **D** 72 m²

3 **a** Sketch the net of a triangular prism with each triangular face having a base of 6 cm and a height of 4 cm and with each rectangular face having a length of 9 cm and a width of 5 cm.

   b Find the surface area of this prism.

4 Find the surface area of each triangular prism.

a
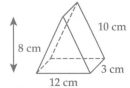
10 cm
8 cm
3 cm
12 cm

b
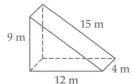
15 m
9 m
12 m
4 m

c
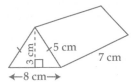
3 cm
5 cm
7 cm
8 cm

d
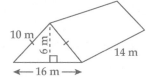
10 m
6 m
14 m
16 m

e
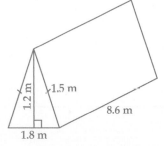
1.2 m
1.5 m
8.6 m
1.8 m

f
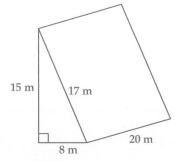
15 m
17 m
8 m
20 m

5 This chocolate bar has the shape of a triangular prism.

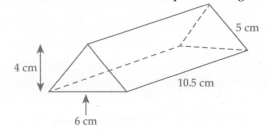
5 cm
4 cm
10.5 cm
6 cm

How many cm² of paper is needed to wrap around the chocolate bar? (Assume there is no overlap.)

ISBN 9780170351058

A cylinder and its net

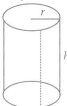 

The rectangle is formed by cutting the cylinder along one edge (see dotted line) and opening it up. The length of this rectangle is $2\pi r$ as it is the circumference of the circle and the width of the rectangle is h, the height of the cylinder.

Area of both circles = $2 \times \pi r^2$

Area of rectangle = $2\pi r \times h$

> **SURFACE AREA OF A CYLINDER**
>
> SA = $2 \times \pi \times$ radius^2 + $2 \times \pi \times$ radius $\times$ height
>
> SA = $2\pi r^2 + 2\pi rh$

This formula is for a **closed** cylinder, where there is a top and a bottom.

An **open** cylinder has a base but no top so its **SA = $\pi r^2 + 2\pi rh$**, only one circle not two.

EXAMPLE 10

Find the surface area of each cylinder, correct to 1 decimal place.

a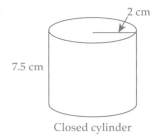

Closed cylinder

b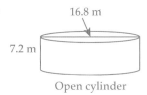

Open cylinder

SOLUTION

a Surface area = $2\pi r^2 + 2\pi rh$

$= 2 \times \pi \times 2^2 + 2 \times \pi \times 2 \times 7.5$

$= 119.3805...$

≈ 119.4 cm^2

b Surface area = $\pi r^2 + 2\pi rh$

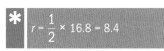

$= \pi \times 8.4^2 + 2 \times \pi \times 8.4 \times 7.2$

$= 601.6778...$

≈ 601.7 m^2

1   What are the shapes of the faces of a closed cylinder? Select the correct answer **A**, **B**, **C** or **D**.

   A 1 circle, 1 rectangle   **B** 2 circles, 1 rectangle
   C 2 circles, 2 rectangles   **D** 1 circle, 2 rectangles

2   Find the surface area of a closed cylinder with radius 3 cm and height 6 cm.
   Select **A**, **B**, **C** or **D**.

   A 169.6 cm²   **B** 141.4 cm²   **C** 113.1 cm²   **D** 358.1 cm²

3   **a**   Draw the net of a cylinder with radius 8 cm and height 5 cm.

   b   Calculate its surface area correct to one decimal place.

4   Find, correct to one decimal place, the surface area of each solid.

a
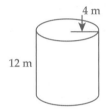
4 m
12 m

b   open cylinder

1.4 m
3.2 m

c   5.3 cm
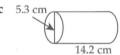
14.2 cm

d
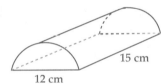
15 cm
12 cm

e   open cylinder
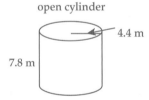
4.4 m
7.8 m

f
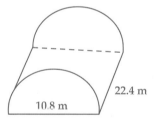
22.4 m
10.8 m

g
3.5
6

5   **a**   The outside of a cylindrical water tank of height 15.6 m and diameter 7.4 m needs
       to be painted. The tank does not have a lid and its base cannot be painted. Find the
       curved surface area of the tank, correct to two decimal places.

   b   If one can of paint can cover 25 m² of the tank, how many full cans of paint are required?

   c   If one can costs $68.50, what is the total cost of the paint required?

volume The amount of space occupied by a solid shape. Volume is measured in cubic units such as mm^3, cm^3 and m^3.

capacity The amount of liquid or material that a container can hold when full. It is measured in millilitres, litres, kilolitres or megalitres.

cross-section The shape of a slice that is cut through a prism, parallel to the ends. A cross-section must be the same shape all along the prism.

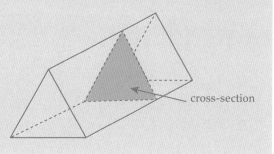

cross-section

Metric units of volume

cubic centimetre (cm^3)

1 cm^3

1 cm

$1 \text{ cm} = 10 \text{ mm}$
$1 \text{ cm}^3 = 1 \text{ cm} \times 1 \text{ cm} \times 1 \text{ cm}$
$= 10 \text{ mm} \times 10 \text{ mm} \times 10 \text{ mm}$
$= 1000 \text{ mm}^3$

cubic metre (m^3)

1 m^3

1 m

$1 \text{ m} = 100 \text{ cm}$
$1 \text{ m}^3 = 1 \text{ m} \times 1 \text{ m} \times 1 \text{ m}$
$= 100 \text{ cm} \times 100 \text{ cm} \times 100 \text{ cm}$
$= 1\,000\,000 \text{ cm}^3$

Volume
$1 \text{ cm}^3 = 1000 \text{ mm}^3$
$1 \text{ m}^3 = 1\,000\,000 \text{ cm}^3$

Capacity
$1 \text{ L} = 1000 \text{ mL}$
$1 \text{ kL} = 1000 \text{ L}$
$1 \text{ ML} = 1000 \text{ kL} = 1\,000\,000 \text{ L}$

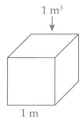

1 mL

1 cm^3

$\times 1\,000\,000 =$

1 cm^3 contains 1 mL

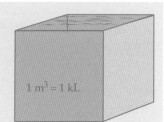

$1 \text{ m}^3 = 1 \text{ kL}$

1 m^3 contains 1 kL or 1000 L

EXAMPLE 11

Convert:

a 4 cm^3 to mm^3 **b** 1.9 m^3 to cm^3 **c** 230 000 mm^3 to cm^3

d 5.4 kL to L   **e** 610 mL to L   **f** 1.35 m^3 to L

SOLUTION

a $4 \text{ cm}^3 = 4 \times 1000 \text{ mm}^3$ ⟵ $1 \text{ cm}^3 = 1000 \text{ mm}^3$
$= 4000 \text{ mm}^3$

b $1.9 \text{ m}^3 = 1.9 \times 1\,000\,000 \text{ cm}^3$ ⟵——— $1 \text{ m}^3 = 1\,000\,000 \text{ cm}^3$
$= 1\,900\,000 \text{ cm}^3$

c $230\,000 \text{ mm}^3 = (230\,000 \div 1000) \text{ cm}^3$ ⟵——— $1 \text{ cm}^3 = 1000 \text{ mm}^3$
$= 230 \text{ cm}^3$

d $5.4 \text{ kL} = 5.4 \times 1000 \text{ L}$ ⟵——— $1 \text{ kL} = 1000 \text{ L}$
$= 5400 \text{ L}$

e $610 \text{ mL} = (610 \div 1000) \text{ mL}$ ⟵——— $1 \text{ L} = 1000 \text{ mL}$
$= 0.61 \text{ L}$

f $1.35 \text{ m}^3 = 1.35 \times 1000 \text{ L}$ ⟵——— $1 \text{ m}^3 = 1000 \text{ L}$
$= 1350 \text{ L}$

VOLUME OF A PRISM

V = area of base or cross-section × height

$V = Ah$

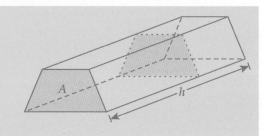

EXAMPLE 12

Find the volume of each prism.

a

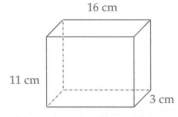

16 cm

11 cm

3 cm

b

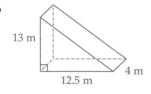

13 m

12.5 m

4 m

SOLUTION

a Volume $= 16 \times 3 \times 11$ ⟵——— Volume = length × width × height for a rectangular prism
$= 528 \text{ cm}^3$

b Find A, the area of the base or cross-section first.

$A = \dfrac{1}{2} \times 12.5 \times 13$ ⟵——— Area of a triangle

$= 81.25 \text{ m}^2$

$V = Ah$

$= 81.25 \times 4$ ⟵——— $h = 4$

$= 325 \text{ m}^3$

1 Convert 8.45 ML to L. Select the correct answer **A, B, C** or **D**.

  A 845 000   **B** 84 500   **C** 8450   **D** 8 450 000

2 Convert 7.6 cm^3 to mm^3. Select **A, B, C** or **D**.

  A 7 600 000   **B** 7600   **C** 76 000   **D** 760 000

3 Copy and complete each equation.

  a 300 L = _____ mL   **b** 22 mm^3 = _____ cm^3

  c 8.6 L = _____ kL   **d** 23.5 mL = _____ cm^3

  e 6840 L = _____ m^3 **f** 52 ML = _____ L

  g 7.8 m^3 = _____ cm^3 **h** 76.4 kL = _____ m^3

  i 44 700 m^3 = _____ L   **j** 240 L _____ kL

  k 0.8 cm^3 = _____ mm^3 **l** 12.8 m^3 = _____ cm^3

4 Find the volume of each prism.

a
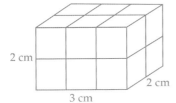

b
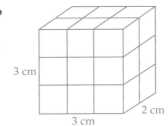

c

d
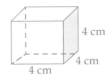

e
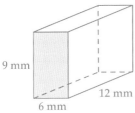

f

g
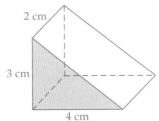

h
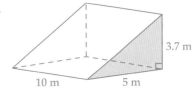

5 A gift box in the shape of a rectangular prism is 9 cm long, 6 cm wide and 15 cm high. Find its capacity in:

  a cubic centimetres   **b** millilitres   **c** litres.

VOLUME OF A PRISM

V = area of base or cross-section × height

$V = Ah$

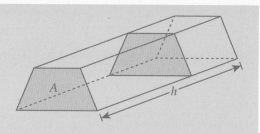

VOLUME OF A CYLINDER

V = area of circular base × height

V = π × radius² × height

$V = \pi r^2 h$

EXAMPLE 13

Find the volume of each prism.

a

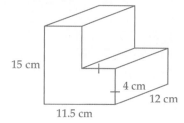

15 cm
4 cm
12 cm
11.5 cm

b

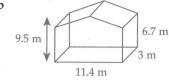

9.5 m
6.7 m
3 m
11.4 m

SOLUTION

a $A = (7.5 \times 15) + (4 \times 4)$ ⟵ Area of L-shape

$= 128.5 \text{ cm}^2$

$V = Ah$

$= 128.5 \times 12$ ⟵ $h = 12$

$= 1542 \text{ cm}^3$

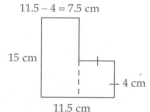

11.5 − 4 = 7.5 cm
15 cm
4 cm
11.5 cm

b $A = \left(\dfrac{1}{2} \times 11.4 \times 2.8\right) + (11.4 \times 6.7)$ ⟵ Area of a triangle + rectangle

✳ Triangle's height = 9.5 − 6.7 = 2.8 m

$= 92.34 \text{ m}^2$

$V = Ah$

$= 92.34 \times 3$ ⟵ $h = 3$

$= 277.02 \text{ m}^3$

Volumes of prisms and cylinders

EXAMPLE 14

Find, correct to 1 decimal place, the volume of each cylinder.

a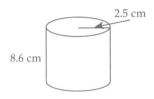

b

SOLUTION

a $V = \pi r^2 h$

$= \pi \times 2.5^2 \times 8.6$

$= 168.8606...$

$\approx 168.9 \text{ cm}^3$

b $V = \pi r^2 h$

$= \pi \times 6.3^2 \times 6.4$ ⟵ $r = \dfrac{1}{2} \times 12.6 = 6.3$

$= 798.0147...$

$\approx 798.0 \text{ m}^3$

EXERCISE 11-08

1 Find the volume of a prism with a base of area 18.4 m² and height 15 m. Select the correct answer **A**, **B**, **C** or **D**.

A 138 m³   **B** 867.1 m³   **C** 276 m³   **D** 92 m³

2 Find the volume of a cylinder with a diameter of 16 cm and a height of 5.2 cm. Select **A**, **B**, **C** or **D**.

A 4182.1 cm³   **B** 1045.5 cm³   **C** 522.8 cm³   **D** 2091 cm³

3 Name each prism and sketch its cross-section.

a   **b**    **c**

4 **a** What shapes make up the cross-section of this prism?

b Find the area A of the cross-section.

c For this prism, what is the value of h in the formula $V = Ah$?

d Find the volume of this prism.

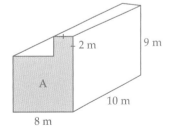

5 Find the volume of each prism.

a

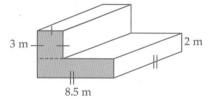

b

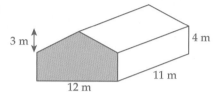

c

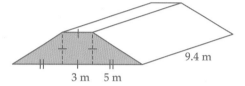

d

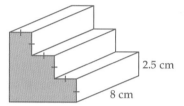

6 Find, correct to 1 decimal place, the volume of each cylinder. All measurements are in cm.

a

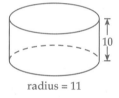

b

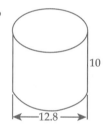

c

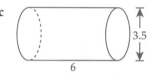

7 A country cottage has a water tank as shown.

   a What is its volume, correct to three decimal places?

   b How many litres of water can it hold? Answer to the nearest litre.

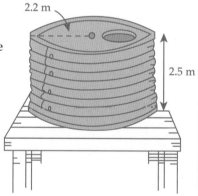

8 A large fish tank in a restaurant has this shape.

   a Calculate the volume of the fish tank.

   b Calculate the capacity of the fish tank in litres.

   c How many fish can be placed in the fish tank if each fish needs 25 L of water space?

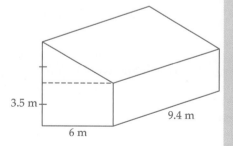

CODE PUZZLE

Use this table to decode the words used in this chapter.

A	B	C	D	E	F	G	H	I	J	K	L	M
1	2	3	4	5	6	7	8	9	10	11	12	13

N	O	P	Q	R	S	T	U	V	W	X	Y	Z
14	15	16	17	18	19	20	21	22	23	24	25	26

1 11 – 9 – 20 – 5
2 18 – 8 – 15 – 13 – 2 – 21 – 19
3 16 – 9
4 17 – 21 – 1 – 4 – 18 – 1 – 14 – 20
5 19 – 5 – 13 – 9 – 3 – 9 – 18 – 3 – 12 – 5
6 20 – 18 – 1 – 16 – 5 – 26 – 9 – 21 – 13
7 4 – 9 – 1 – 7 – 15 – 14 – 1 – 12
8 1 – 18 – 5 – 1
9 16 – 1 – 18 – 1 – 12 – 12 – 5 – 12 – 15 – 7 – 18 – 1 – 13
10 3 – 15 – 13 – 16 – 15 – 19 – 9 – 20 – 5
11 16 – 18 – 9 – 19 – 13
12 19 – 21 – 18 – 6 – 1 – 3 – 5 1 – 18 – 5 – 1
13 3 – 25 – 12 – 9 – 14 – 4 – 5 – 18
14 22 – 15 – 12 – 21 – 13 – 5
15 3 – 1 – 16 – 1 – 3 – 9 – 20 – 25

PRACTICE TEST 11

Part A General topics

Calculators are not allowed.

1 Round 0.078 92 to two significant figures.

2 Does the point $(0, -4)$ lie on the x-axis or y-axis?

3 Find the value of $3x^2$ if $x = -2$.

4 Write 476 000 in scientific notation.

5 Find y.

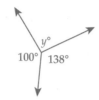

6 James and Samantha share a prize of $2000 in the ratio 2 : 3. What is James' share?

7 Simplify $2h^2 \times 3h^6$

8 Evaluate $30 - 6 \times 4 + 20$

9 Find the mode of this set of data.

Stem	Leaf
1	1 4 4 5
2	0 3 6
3	2 4
4	7

10 Find the median of the data in Question 9.

Part B Area and volume

Calculators are allowed.

11–01 Area

11 Find the area of a rectangle with length 12 cm and width 4.5 cm. Select the correct answer **A**, **B**, **C** or **D**.

   A 54 cm²   **B** 108 cm²   **C** 33 cm²   **D** 27 cm²

12 Find the area of this triangle. Select **A**, **B**, **C** or **D**.

   A 122.64 m²   **B** 245.28 m²

   C 61.32 m²   **D** 46 m²

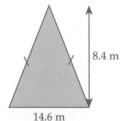

14.6 m

8.4 m

11–02 Area of a circle

13 Find the area of a circle with radius 6.8 m. Select **A**, **B**, **C** or **D**.

   A 42.7 m²   **B** 145.3 m²   **C** 72.7 m²   **D** 21.4 m²

14 Find, correct to one decimal place, the area of each shape.

a

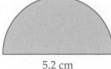

5.2 cm

b

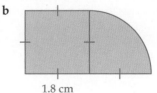

1.8 cm

11-03 Areas of trapeziums, kites and rhombuses

15 Find the area of each quadrilateral.

a
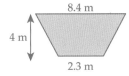
8.4 m
4 m
2.3 m

b
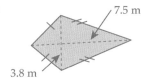
7.5 m
3.8 m

16 What is the area of a rhombus with diagonals 22.4 and 16.3 cm?

11-04 Surface area of a rectangular prism

17 Find the surface area of each prism.

a

1.5 cm

b

3.9 m
1.6 m
9.4 m

11-05 Surface area of a triangular prism

18 Find the surface area of this prism.

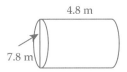
5 cm
3 cm
14 cm
8 cm

11-06 Surface area of a cylinder

19 Find, correct to 1 decimal place, the surface area of this cylinder.

4.8 m
7.8 m

11-07 Volume

20 Copy and complete each equation.

a $9.2 \text{ m}^3 = \underline{\hspace{1cm}} \text{cm}^3$

b $275 \text{ L} = \underline{\hspace{1cm}} \text{kL}$

c $85 \text{ m}^3 = \underline{\hspace{1cm}} \text{L}$

11-08 Volumes of prisms and cylinders

21 Find the volume of each solid. Write the answer to part **b** correct to one decimal place.

a
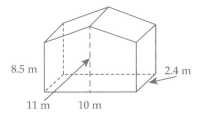
8.5 m
2.4 m
11 m   10 m

b
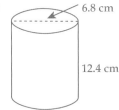
6.8 cm
12.4 cm

Shutterstock.com/Christian Mueller

INVESTIGATING DATA

12

IN THIS CHAPTER YOU WILL:

- know the difference between a sample and a census
- find the mean, mode, median and range of a set of numerical data
- draw dot plots and stem-and-leaf plots
- complete a frequency distribution table
- draw frequency histograms and polygons
- identify the shape of a frequency distribution
- compare data on a back-to-back stem-and-leaf plot
- find the interquartile range of a set of numerical data
- draw and interpret box-and-whisker plots

shutterstock.com/suphakit73

WORDBANK

data A collection of raw facts or information.

population In statistics, all of the items under investigation. For example, if a survey is done on the cattle on a farm, the population is all cattle on this farm.

sample A part of the population selected to be surveyed.

census A survey of the entire population.

A **census** surveys the whole population but can be expensive and time-consuming to conduct.
A **sample** surveys part of the population and should:
- represent the population fairly
- be **random**, where every item of the population has an equal chance of being selected
- be **not biased**, where one type of item is not favoured over another
- be **large**, because the larger the sample, the more accurately it represents the population

EXAMPLE 1

Determine whether a sample or a census would be more suitable for each survey.

a Finding the most popular TV program in Australia.

b Collecting data on the number of children in childcare in Tasmania.

c Surveying people's opinions on the sale of cigarettes.

SOLUTION

a Sample, as it would be too expensive to conduct a census of all Australians for this reason.

b Census, as reliable data should be available at each childcare centre in Tasmania.

c Sample, as the population is too large and it would be too time-consuming to survey everyone.

iStockphoto.com/sdominick

1   Which word best describes what a sample should be? Select the correct answer **A**, **B**, **C** or **D**.

   A expensive   **B** random   **C** biased   **D** time-consuming

2   Is each statement true or false?

   a   A census surveys the entire population.

   b   A sample is always an accurate predictor of future trends.

   c   A census is always an accurate predictor of future trends.

   d   A census can be biased.

   e   A sample should represent the population.

   f   A census requires many people to organise and run.

3   State whether a sample or a census would be more suitable for each survey.
   Give a reason for your answer.

   a   Finding the number of students enrolled in a high school.

   b   Finding Australia's most favourite talent show on TV.

   c   Surveying shoppers on the most popular brand of ice-cream.

   d   Testing the quality of shoes made in a factory.

   e   Finding the population of Canberra.

   f   Surveying the best pizza restaurant in Melbourne.

   g   Testing the speed of all competitors in a swimming race.

   h   Finding the best-selling book in Australia last month.

4   State whether the following methods for selecting a sample of 20 people are random or
   biased.

   a   Giving a number to each person, shuffling the numbers in a bowl and drawing out
       20 numbers.

   b   Selecting the last 20 customers in the store before it closes.

   c   Selecting 20 names from different parts of a spreadsheet that lists all the students in
       your school.

   d   Choosing 20 students from the 10A class when surveying students about canteen food.

   e   Interviewing 20 customers from McDonalds when investigating teenagers' favourite
       foods.

5   The table below shows the number of students in each year at Nelson High School.

Year	7	8	9	10	11	12
Students	130	112	108	135	143	122

   a   What is the school population?

   b   What percentage of the school population are Year 10 students?

   c   If I wish to survey a sample of 50 Nelson High School students about their sports
       interests, how many students should I survey from Year 10?

6   Investigate how often a national census is held in Australia. When will the next census be?

12–02 | The mean, mode, median and range

WORDBANK

outlier An extreme score that is much higher or much lower than the other scores in the data set. An outlier affects the value of the mean.

The mean, mode and median are called **measures of location** because they measure the central or middle position of a set of data.

Mean $= \bar{x} = \dfrac{\text{sum of scores}}{\text{number of scores}}$

Mode = the most popular score(s).

For the **median**:
- order the scores from lowest to highest
- if there are an odd number of scores, **median = middle score**
- if there are an even number of scores, **median = average of the two middle scores**.

Range = Highest score – lowest score

EXAMPLE 2

For this set of data:

8　7　5　8　9　7　19　6　8　7　9　8

find:

a the mean correct to one decimal place

b the mode　　**c** the median　　**d** the range　　**e** any outliers

SOLUTION

a Mean, $\bar{x} = \dfrac{8+7+5+8+9+7+19+6+8+7+9+8}{12}$ ← $\dfrac{\text{sum of scores}}{\text{number of scores}}$

$= \dfrac{101}{12}$

$= 8.4166\ldots$

≈ 8.4

b Mode = 8 ←——— most popular score

c For the median, list the scores in ascending order.

5　6　7　7　7　⑧　⑧　8　8　9　9　19

✱ As there is an even number of scores, there are 2 scores in the middle.

Median $= \dfrac{8+8}{2}$ ←——— average of the 2 middle scores

$= 8$

✱ Note that the mean, mode and median are around the centre of the set of scores.

d Range = 19 – 5 ←——— highest score – lowest score

$= 14$

e Outlier = 19 ←——— 19 is much higher than all the other scores.

Developmental Mathematics Book 4　　　　ISBN 9780170351058

1 Is each statement true or false?

  a The most popular score is called the mode.

  b The median is the difference between the highest and lowest scores.

  c An outlier is close to the median.

  d The value of the mean depends on every score in the set of data.

  e If there is an odd number of scores, then there are two middle scores.

  f An outlier can affect the mode of a set of data.

2 For each set of data, find the mode, range, median, mean (one decimal place) and any outliers.

a	3	5	4	3	6	8	7	5	3	9
b	10	11	14	10	15	30	13	14	12	
c	8	10	8	9	11	12	8	9		
d	21	23	25	21	8	24	21	25	21	
e	55	54	56	52	38	55	52	53	55	56
f	100	101	105	101	103	100	101	104	102	

3 Seven students were surveyed on how much cash they carried.

  $15 $18 $19 $21 $45 $16 $15

  a Find the mean, correct to the nearest cent.

  b What is the outlier?

  c Calculate the mean without the outlier.

  d How does the outlier affect the mean?

4 Three teams compete at a school swimming carnival and the times (in seconds) the swimmers took to swim a lap of the pool are listed.

  Reds: 46.6   42.7   45.5   48.1   51.6   42.8

  Greens: 46.7   51.4   48.6   47.2   52.4   41.2

  Blues: 47.9   55.8   56.4   48.8   51.6   53.7

  a Find the median and range for each team's times.

  b The team to represent the school at the zone carnival is the one with the lowest median. Which team is this?

  c Which team had the fastest swimmer?

5 Last week, eight houses were sold in Cypress Crescent. The selling prices were:

$890 000	$845 000	$724 500	$890 000
$982 000	$1 450 800	$725 250	$621 000

  a What is the mode?

  b Find the mean and median.

  c What is the outlier?

  d Which measure does the outlier affect the most?

WORDBANK

dot plot A diagram showing frequency of data scores using dots.

stem-and-leaf plot A table listing data scores, where the tens or hundreds digits are in the stem and the units digits are in the leaf.

cluster Where scores in a set of data are close together in a group.

EXAMPLE 3

This dot plot shows the daily sales, rounded to the nearest $50, from a newsagent for 2 weeks.

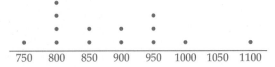

Find:

a the range

b the mode

c the mean, correct to the nearest cent

d the median

e any clusters or outliers

SOLUTION

a Range $= \$1100 - \$750$ ⟵ highest score – lowest score

$\qquad = \$350$

b Mode $= \$800$ ⟵ most common score

c Mean $= \dfrac{750 + 4 \times 800 + 2 \times 850 + 2 \times 900 + 3 \times 950 + 1000 + 1100}{14}$ ⟵ $\dfrac{\text{sum of scores}}{\text{no. of scores (dots)}}$

$\qquad = \$885.7142...$

$\qquad \approx \$885.71$

d Median $= \dfrac{\$850 + \$900}{2}$ ⟵ 14 scores: the two middle dots (7th and 8th) are 850 and 900

$\qquad = \$875$

e A cluster of scores occurs from $800 to $950.

There is one outlier of $1100.

EXAMPLE 4

This stem-and-leaf plot shows the number of people visiting a small cinema each day for 18 days.

Stem	Leaf
3	3 5 8 8
4	2 5 5 6 8
5	4 6 9
6	2 5 7 8
7	1 2

Find:

a the mode

b the median

c the mean, correct to one decimal place

d the range

SOLUTION

For a stem-and-leaf plot, the stem is the tens digit and the leaf is the units digits, so the scores are 33, 35, 38, 38, up to 72.

a Mode = 38 and 45 ⟵ Both appear twice.

b Median = $\dfrac{48+54}{2}$ ⟵ 18 scores: the two middle scores (9th and 10th) are 48 and 54.

   = 51

c Mean = $\dfrac{33+35+38+...+72}{18}$ ⟵ $\dfrac{\text{sum of scores}}{\text{no. of scores}}$

   = $\dfrac{944}{18}$

   = 52.4444...

   ≈ 52.4

d Range = 72 − 33 ⟵ highest score − lowest score

   = 39

1 In a dot plot, the score with the highest column of dots is what? Select the correct answer A, B, C or D.

A the range   B the mean   C the median   D the mode

2 In a stem-and-leaf plot, the stem usually shows the:

A mode   B median   C units digit   D tens digit

3 Copy and complete this paragraph.

A dot plot is a diagram that shows the scores using a _____ axis and a _____ to represent each score. A stem-and-leaf plot lists each _____ by using the tens or _____ digit as the stem and the units digit as the _____.

4 Draw a dot plot for each set of scores.

a	3	4	6	7	8	4	8	5	4	6	8
b	10	12	14	13	10	11	12	14	11	11	12
c	21	23	21	25	24	23	22	23	24	30	23
d	50	55	54	51	55	54	53	54	50	54	52
e	16	18	17	16	18	17	10	11	18	15	16

5 For each dot plot drawn in Question 4, find:

i the mode   ii the median   iii the range.

6 a Which dot plot drawn in Question 4 shows a cluster of scores?

b Which dot plot has an outlier?

7 The data below shows the daily number of people visiting the local swimming pool.

46   72   83   66   66   42   75   61   73

48   84   71   66   75   58   62   54   80

a Draw an ordered stem-and-leaf plot for this data.

b What is the mode?

c Find the median.

d What is the mean, correct to one decimal place?

8 The data below shows the daily number of people visiting *The Modal Class* restaurant for a seafood buffet.

38   72   97   42   35   54   65   74   48   39   43

72   68   72   55   37   56   67   72   64   75

a Draw a stem-and-leaf plot displaying this data.

b Find the range and mode for this data.

c Calculate the mean daily sales takings for the restaurant if each person pays $45.

d What is the median?

e Are there any clusters of scores? If so, where are they?

f Is there an outlier? If so, what is it?

frequency The number of times a score occurs in a data set.

cumulative frequency A running total of frequencies used for finding the median, combining the frequencies of all scores less than or equal to the given score.

frequency table A table that shows the frequency of each score in a data set.

For data in a frequency table:
- a cumulative frequency column can be included to find the median
- an fx column can be included to find the mean
- mean: $\overline{x} = \dfrac{\text{sum of } fx}{\text{sum of } f}$

EXAMPLE 5

A group of students were surveyed on the number of pets they have at home.

2	3	1	3	0	2	1	1	0	5	2
3	2	4	2	1	3	5	4	1	3	2

a Arrange these scores in a frequency table, including columns for cumulative frequency and fx.

b For this data set, find:

   i the range   **ii** the mode

   iii the median   **iv** the mean, correct to two decimal places

SOLUTION

a

Score (x)	Tally	Frequency (f)	Cumulative frequency	fx
0	||	2	2	0
1	⊔⊓⊓	5	7	5
2	⊔⊓⊓ |	6	13	12
3	⊔⊓⊓	5	18	15
4	||	2	20	8
5	||	2	22	10
	Totals	22		50

* Cumulative frequency is a running total of frequencies. fx means 'frequency (f) × score (x)'.

b **i** Range = 5 − 0   ⟵——— Highest score − lowest score (from the Score column).

        = 5

   ii Mode = 2   ⟵——— The score with the highest frequency (6).

iii There are 22 scores, so the median is the average of the 11th and 12th scores.

From the cumulative frequency column, it can be seen that the 8th to 13th scores are all 2s, so the 11th and 12th scores are both 2.

$$\text{Median} = \frac{2+2}{2}$$

$$= 2$$

iv Mean $\bar{x} = \dfrac{\text{sum of } fx}{\text{sum of } f}$

$$= \frac{50}{22} \longleftarrow \text{Use the totals at the bottom of the table.}$$

$$= 2.2727\ldots$$

$$\approx 2.27$$

EXERCISE 12-04

1. Which column do you use to calculate the cumulative frequency in a frequency table? Select the correct answer **A**, **B**, **C** or **D**.

   A score　　　　　**B** tally　　　　　**C** *fx*　　　　　**D** frequency

2. Which column do you use to find the median in a frequency table? Select **A**, **B**, **C** or **D**.

   A score　　　**B** frequency　　　**C** *fx*　　　**D** cumulative frequency

3. Copy and complete this frequency table, then use it to find the range, mode and median of this set of data.

Score	Frequency	Cumulative frequency
30	3	
31	6	
32	8	
33	10	
34	12	
35	4	
36	2	

4   The number of homes sold each month in a coastal region are listed below.

21   25   30   29   22   28   30   24   22   23   26   22
27   29   30   22   24   22   29   28   22   30   21   22

   a   Arrange these scores in a frequency table, including columns for cumulative frequency and *fx*.

   b   What is the range for this data?

   c   What is the most likely number of homes sold each month?

   d   Calculate the mean, correct to two decimal places.

   e   Find the median.

5   The weekly wages (in dollars) of a group of painters are listed below.

490   520   580   550   580   540   480   520   560   570
540   570   540   520   530   590   580   510   560   580
520   540   550   540   560   590   550   530   540   570

   a   Arrange these values in a frequency table, including a column for cumulative frequency.

   b   What is the highest weekly wage?

   c   Find the range of weekly wages.

   d   Ken, a new painter, is told that he will earn the most common of the weekly wages. What will his wage be?

   e   What is the median weekly wage?

6   A sample of people was surveyed on how often they visited a doctor in one year.

   2   3   0   2   3   4   6   5   3   1
   6   0   3   1   4   7   2   2   4   5
   8   4   2   3   5   4   3   2   2   1

   a   Complete a frequency table for this data set and find:

   i   the range   ii   the mode

   iii   the median   iv   the mean (correct to two decimal places)

   b   How much does the average person spend on doctors and medicine in a year, at $52 per visit, correct to the nearest dollar?

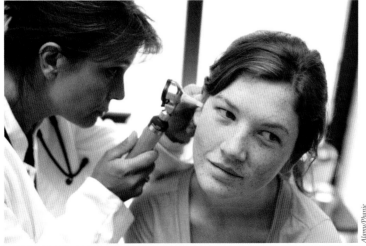

Alamy/Phanie

WORDBANK

frequency histogram A column graph that shows the frequency of each score. The columns are joined together.

frequency polygon A line graph that shows the frequency of each score, drawn by joining the middle of the top of each column in a histogram.

EXAMPLE 6

The frequency histogram and polygon below show the ages of the children at a birthday party.

Frequency histogram

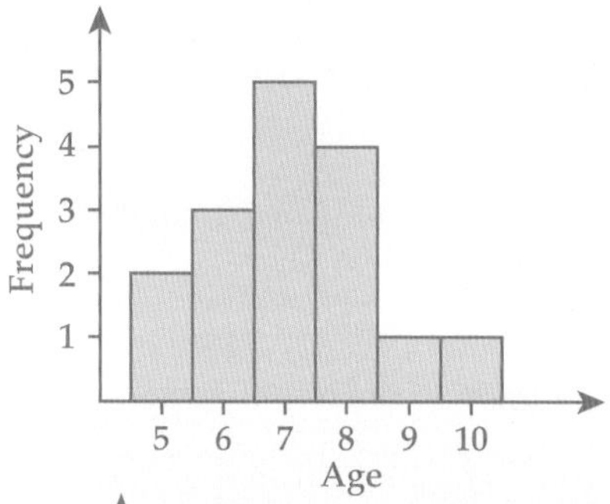

Frequency polygon

✱ The columns are centred on the scores (ages), leaving a 'half-column gap' on the left.

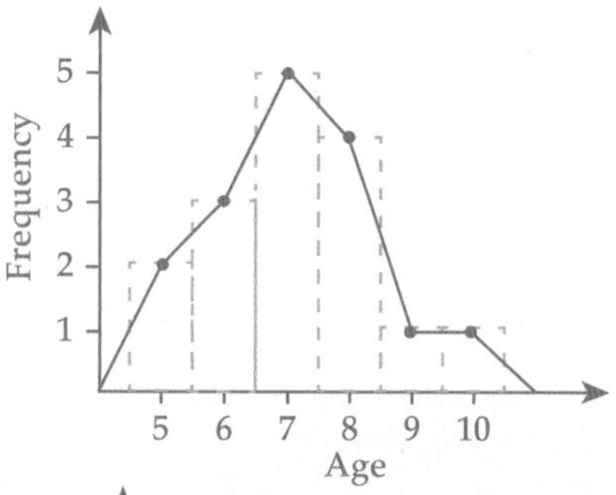

✱ The polygon starts and ends on the horizontal axis, and joins the top of each column.

For this data, find:

a the range

b the mode

c the mean, correct to one decimal place

SOLUTION

a Range = 10 − 5 ⟵ Highest age – lowest age

      = 5

b Mode = 7 ⟵ Age with the highest column.

c To calculate the mean, use the histogram or polygon to complete a frequency table with an *fx* column.

x	f	fx
5	2	10
6	3	18
7	5	35
8	4	32
9	1	9
10	1	10
Totals	16	114

$$\text{Mean} = \frac{114}{16} \quad \longleftarrow \quad \frac{\text{sum of } fx}{\text{sum of } f}$$

$$= 7.125$$

$$\approx 7.1$$

1 Which statement is true about the columns in a frequency histogram? Select the correct answer **A**, **B**, **C** or **D**.

   A They have equal heights   **B** The last column is always the highest

   C They have different widths   **D** There are no gaps between them

2 Which statement is true about a frequency polygon? Select **A**, **B**, **C** or **D**.

   A It is a special type of column graph

   B It joins the corners of each column of the histogram

   C It starts and ends on the horizontal axis

   D It is similar to a dot plot

3 Copy and complete this paragraph.

   A frequency histogram is a type of _____ graph, where the columns are _____ together and are of _____ width. A frequency polygon is a type of _____ graph, where the points are the middle of the top of each _____ in the histogram and joined by _____.

4 A class of students were surveyed on the number of text messages sent yesterday.

12	16	18	13	15	15	14	16	15	14	18
17	13	12	15	14	15	17	13	15	13	15

   a Complete a frequency table, then draw a frequency histogram and polygon for the data.

   b Find the range and mode for this data.

   c Explain the meaning of the range and mode for this data set.

5 Find the range, mode and mean (correct to two decimal places) for each graph.

   a
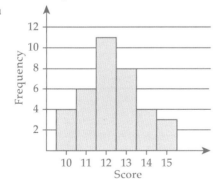

   b
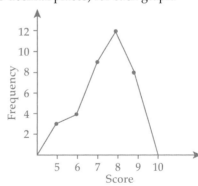

6 The marks below are the yearly exam results for a Year 10 maths class.

74	75	79	74	72	72	76	78	73	76
75	73	76	77	78	74	76	72	74	74

   a Draw a frequency table of the marks.

   b Graph the results on a frequency histogram and polygon.

   c Find the range and mode of the data.

   d Calculate the mean mark.

   e What percentage of students scored above the mean?

WORDBANK

symmetrical When the scores are evenly spread or balanced around the centre.

skewed When the scores are bunched or clustered at one end, while the other end has a tail.

bimodal (pronounced 'bye-mode-l') When the scores have two peaks.

When looking at histograms, dot plots and stem-and-leaf plots, the shape of the data can be seen by drawing a curve around the graph.

Symmetrical
The scores are clustered around the centre.

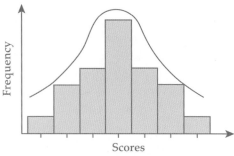

Positively skewed
The scores are clustered to the left and the tail is to the right.

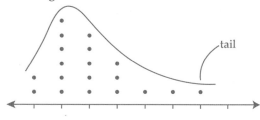

Negatively skewed
The scores are clustered to the right and the tail is to the left.

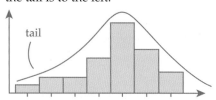

Bimodal
Two peaks

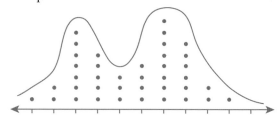

EXAMPLE 7

For each data display:

i   describe the shape

ii   identify any clusters or outliers

a

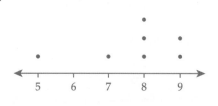

b

Stem	Leaf		
4	2	3	6
5	1	2	
6	2		
7	3		

✱ When looking at the shape of a stem-and-leaf plot, turn the plot sideways so that the stem is at the bottom.

SOLUTION

a i The data is negatively skewed.
ii 5 is an outlier and there is clustering at 8 and 9.

b i The data is positively skewed
ii There is a cluster at the 40s and 50s stems but no outliers.

Developmental Mathematics Book 4

ISBN 9780170351058

1 Which phrase is true about a negatively-skewed graph? Select the correct answer **A**, **B**, **C** or **D**.

    A The scores are clustered to the left   **B** The tail points to the left

    C There are two peaks   **D** The peak is on the left

2 For each data display:

    i describe the shape   **ii** identify any clusters or outliers

a
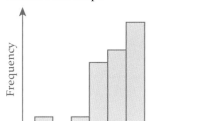

b

c

Stem	Leaf		
12	1		
13	2	8	
14	3	5	7
16	4	5	9

d
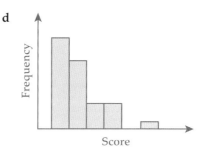

3 This stem-and-leaf plot shows the monthly maximum temperatures in Mt Gotham for 15 months.

Stem	Leaf			
1	0	3	3	3
1	6	8	9	9
2	3	5		
2	6	8		
3	1	3		
3	8			

    a Describe the shape of the graph.

    b Find the mode and range.

    c Calculate the mean temperature correct to one decimal place.

    d What is the median temperature?

    e Where are most of the scores clustered?

4 If 38 is considered an outlier for the scores in Question **3**, which measure of location does it affect the most? Re-calculate the mode, mean and median without using the outlier to answer this question.

5 Sketch an example of a symmetrical graph.

Back-to-back stem-and-leaf **plots** compare two data sets.

EXAMPLE 8

This back-to-back stem-and-leaf plot compares the project marks of two Science classes –
10 Green and 10 Yellow.

10 Green			10 Yellow
		5	4 4 8
	9	6	5 7 8 8
8 5 2		7	2 4 7 8
6 6 5 4 2		8	4
8 3 1		9	

* ■ The two stem-and-leaf plots share the same stem.
  ■ 10 Yellow's marks are on the right, and are ordered left-to-right.
  ■ 10 Green's marks are on the left, and are ordered right-to-left.

a Which class scored the highest mark? What was the mark?

b Which class has the higher median?

c Which class has the higher range?

d Which class performed better on the project? How does the stem-and-leaf plot show this?

SOLUTION

a 10 Green scored the highest mark, 98.

b Median for 10 Green = $\dfrac{84 + 85}{2}$ = 84.5, Median for 10 Yellow = $\dfrac{68 + 68}{2}$ = 68.

10 Green has the higher median.

c Range for 10 Green = 98 – 69 = 29, Range for 10 Yellow = 84 – 54 = 30.

10 Yellow has the higher range.

d 10 Green performed better on the project. It has the higher median and the stem-and-leaf plot
shows that its scores are higher (further down the stem).

iStockphoto.com/KM6064

1 What is a back-to-back stem-and-leaf plot used for? Select the correct answer **A**, **B**, **C** or **D**.

   A To display a large set of data   **B** To find the mode and range of a set of data

   C To find the median of a set of data   **D** To compare two sets of data

2 Which one of these cannot be seen on a back-to-back stem-and-leaf plot?
   Select **A**, **B**, **C** or **D**.

   A clusters   **B** outliers   **C** the mean   **D** the mode

3 This back-to-back stem-and-leaf plot shows the cricket batting scores of Josie and Ben.

Josie				Stem		Ben		
9	8	5	1	1				
	8	5	3	2	4	6	8	
		6	2	1 3	2	6	6	8
			2	4	2	4	5	6

   a Calculate the range of each batter.

   b By looking at the stem-and-leaf plot, who do you think performed better?

   c Calculate, correct to one decimal place, the mean of each batter.

   d Who performed better according to their mean scores?

4 The daily number of burgers sold by McDavids and Hungry Jim's are listed below.

   McDavids: 342   280   295   314   327   345   299   307   318   326   317

   Hungry Jim's: 306   348   359   367   298   356   345   355   366   354   342

   Use a stem of 28, 29, up to 36 to draw a back-to-back stem-and-leaf plot of the data.

   a Which store sold the most burgers in one day? What was this number?

   b Calculate the range for McDavids' scores.

   c Find which store has the higher median.

   d What is the mean number of burgers (correct to two decimal places) sold for each store?

   e Which store was better at selling burgers?

5 Draw a back-to-back stem-and-leaf plot for the swimming times (in seconds) of two
   competitors in the 100 m race.

   Jess: 92   89   95   86   93   78   99   84

   Vicki: 86   88   74   78   93   94   83   69

   a Find the range of times for each swimmer.

   b Who swam the best time and what was it?

   c Find the median time for each swimmer.

   d Which swimmer had the lower median? Is this the same person who swam the best time?

   e Who is the better swimmer? Give a reason.

6 Do boys send more text messages than girls? Conduct a survey of how many text
   messages each person in your class sent yesterday and show that data on a back-to-back
   stem-and-leaf plot with headings for boys and girls. Calculate the median and mode for
   each group and compare.

WORDBANK

lower quartile The first quartile Q_1, the value that separates the lower $\frac{1}{4}$ of scores.

upper quartile The third quartile Q_3, the value that separates the upper $\frac{1}{4}$ of scores.

interquartile range The difference between the upper and lower quartiles.

The median is the middle score that divides a set of data into two equal parts (halves). Quartiles are the values Q_1, Q_2 and Q_3 that divide a set of data into four equal parts (quarters).

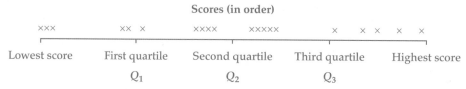

The **range** measures the spread of a set of data but its value is affected by outliers. A better measure of spread is the **interquartile range (IQR)**, which is the difference between Q_3 and Q_1 and not affected by outliers.

To find quartiles and the interquartile range:
- order the scores from lowest to highest and find the median, which is also Q_2
- find the median of the lower half of scores and call it Q_1, the lower quartile
- find the median of the upper half of scores and call it Q_3, the upper quartile
- **Interquartile range (IQR)** = $Q_3 - Q_1$

The interquartile range focuses on the middle 50% of scores and is unaffected by outliers.

EXAMPLE 9

Find the interquartile range for each set of data.

a 12   10   14   12   13   11   15   18   10   12   15

b 23   20   21   23   25   27   28   24   22   27

SOLUTION

a 10   10   (11)   12   12   (12)   13   14   (15)   15   18

            Q_1 Median Q_3

> ✱ Sort the scores from lowest to highest.

Interquartile range = 15 – 11 $Q_3 - Q_1$

                    = 4

b 20   21   (22)   23   23   24   25   (27)   27   28

            Q_1 Median Q_3

IQR = 27 – 22

     = 5

EXAMPLE 10

Find the interquartile range for this stem-and-leaf plot.

Stem	Leaf
4	2 3 6 7
5	1 4 7 8
6	2 4 5 7 9 9
7	1 3 5
8	2 4 6

SOLUTION

$\text{Median} = \dfrac{64 + 65}{2} = 64.5$ ←———— marked in red

$Q_1 = \dfrac{51 + 54}{2} = 52.5, \; Q_3 = \dfrac{71 + 73}{2} = 72$ ←———— marked in blue

$\text{IQR} = 72 - 52.5$

     $= 19.5$

EXERCISE 12-08

1   Which of the following is a measure of spread? Select the correct answer **A**, **B**, **C** or **D**.

    A mode   **B** range   **C** mean   **D** median

2   What does the interquartile range measure? Select **A**, **B**, **C** or **D**.

    A the upper 25% of scores

    B the difference between the mean and the median

    C the lower 25% of scores

    D the middle 50% of scores

3   For each set of data, find the lower quartile, upper quartile and interquartile range.

    a 15   13   16   17   18   15   14   13   12   10   14

    b 8   9   6   7   8   9   5   4   7   6

    c 21   24   25   23   22   20   19   22   23   24

    d 32   33   36   39   34   32   36   37

    e 102   105   103   106   105   108   101   168   105

4 Find the interquartile range for each set of data.

a

5 6 7 8

b

11 12 13

c

Stem	Leaf		
3	2	4	
4	1	3	5
5	3	5	7
6	2	6	8
7	1	5	

d

Stem	Leaf			
6	1	3	5	
7	2	5	6	8
8	3	6	8	9
9	2	5	6	7
10	3	4	7	

5 Two netball teams scored the following number of goals per match.

The Bluebells:	28	32	36	26	42	78
	38	46	32	45	37	36
The Red Ravens	48	36	48	46	54	38
	56	52	48	45	32	46

a Find the range of each team's scores and state which team's range is higher.

b Why is there a large difference between the ranges?

c Find the interquartile range for each team's scores.

d Why is there a smaller difference between the interquartile ranges?

e Which team is more consistent in their scoring? Explain your answer.

Corbis/Paul Cunningham

A **box-and-whisker plot** or **boxplot** is a diagram that shows the quartiles and highest and lowest scores of a set of data. These values are called the **five-number summary:** the lowest score, the lower quartile, the median, the upper quartile and the highest score.

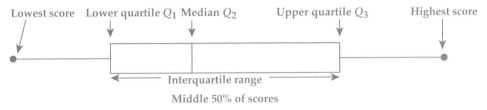

Interquartile range
Middle 50% of scores

EXAMPLE 11

The maximum daily temperature over 11 days in Emu Springs (in °C) is shown here.

15   18   17   16   20   19   16   15   18   20   17

a Find the five-number summary for this data, then represent it on a box-and-whisker plot.

b What is the median daily temperature?

c Find the interquartile range.

d How many scores lie between the lower and upper quartiles?

e What does the end of the right whisker on the box-and-whisker plot represent?

SOLUTION

a Write the scores in ascending order.

Lowest score = 15, Q_1 = 16, Median = 17, Q_3 = 19, highest score = 20.

b Median = 17°C.

c IQR = 19 − 16 = 3

d 4 scores lie between the lower and upper quartiles 16 and 19. ⟵——————— 17, 17, 18 and 18.

e The highest temperature, 20°C.

1   Which measure is NOT shown on a box-and-whisker plot? Select the correct answer **A, B, C** or **D**.
  A the median   **B** the upper quartile
  C the highest score   **D** the mean

2   What does the right edge of the box on a box-and-whisker plot represent? Select **A, B, C** or **D**.
  A the median   **B** the upper quartile
  C the highest score   **D** the mean

3   What are the five values of a five-number summary?

4   Find the five-number summary for this set of data, then represent it on a boxplot.
  5   8   6   7   4   9   5   7   8   9   8

5   Draw box-and-whisker plots for each set of data.
  a 3   5   6   7   8   3   8   5   4   6   7
  b 21   23   24   25   24   23   22   23   24   22   23
  c 50   55   54   51   55   54   53   54   50   54
  d 16   18   17   16   18   17   10   11   18   15   14

6   This box-and-whisker plot shows the number of bread rolls sold at a bakery each day for a month.

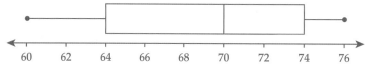

  a What is the median number of bread rolls sold?
  b What are the upper and lower quartiles?
  c Find the interquartile range of the data.
  d What was the greatest number of bread rolls sold in a day?
  e What percentage of the data were between 64 and 74 bread rolls sold?

7   The percentage success rates of two archers during daily target practice are listed.
  Marisa: 82   85   88   83   82   84   86   82   83   84   82
  Stefan: 88   84   86   87   88   86   85   87   86   86   84
  a Represent the data on two box-and-whisker plots using the same scale.
  b What is the range of each archer?
  c Find the interquartile range of each archer.
  d By looking at the box-and-whisker plot, who do you think performed better? Why?
  e What is the median for each archer?
  f Who performed better according to the median?

ISBN 9780170351058

CLUELESS CROSSWORD PUZZLE

Make a copy of this puzzle, then complete the crossword using the words listed below.

BACK-TO-BACK	BIAS	BOX-AND-WHISKER	
CENSUS	DISTRIBUTION	DOT	FREQUENCY
HISTOGRAM	MEAN	MEDIAN	MODE
RANGE	SAMPLE	STEM-AND-LEAF	

Part A General topics

Calculators are not allowed.

1 Evaluate 37×11.

2 Convert 35% to a simple fraction.

3 Write $\frac{2}{3}$, 0.65, 0.6 and 68% in descending order.

4 Simplify $3xy^2 + 4x^2y - xy^2 + 10x^2y$.

5 How many faces does a triangular prism have?

6 Find p.

7 What is the length of the hypotenuse in a right-angled triangle if the other two sides are 8 m and 6 m?

8 A marble is selected from a bag of 2 red, 3 blue and 6 orange marbles. What is the probability that it is not blue?

9 Find the area of this parallelogram.

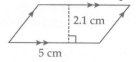

10 How many axes of symmetry has a parallelogram?

Part B Investigating data

Calculators are allowed.

12-01 Sample vs census

11 What is surveyed in a census? Select the correct answer **A**, **B**, **C** or **D**.

  A 10% of the population   **B** all of the population

  C at least half of the population   **D** at least 100 items

12-02 The mean, mode, median and range

12 What is the median of a set of data? Select **A**, **B**, **C** or **D**.

  A the average   **B** highest score – lowest score

  C the middle score   **D** the most popular score

13 For the scores

  36   37   44   38   42   92   34   38   41   35

  find:

  a the mode   **b** the median   **c** any outliers.

14 **a** Calculate the mean of the scores in Question **13**.

  b Does the outlier affect the mean?

12-03 Dot plots and stem-and-leaf plots

15 For the scores

  45   35   40   30   35   40   45   35   45   40

  represent them on:

  a a dot plot

  b a stem-and-leaf plot

12-04 Frequency tables

16 For the data shown in this frequency table, find the:

   a mode   **b** range   **c** median   **d** mean

Score	Frequency
3	4
4	12
5	8
6	8

12-05 Frequency histograms and polygons

17 Graph the data from Question **16** on a frequency polygon.

12-06 The shape of a distribution

18 Describe the shape of each graph below.

   a

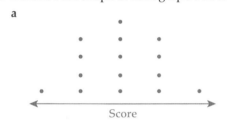

Score

   b

Stem	Leaf
4	3
5	2 6
6	4 5
7	3 4 7
8	1 6 8 9

12-07 Back-to-back stem and leaf plots

19 The waiting times (in minutes) for patients at two doctors' offices are shown. Find the range and median for each doctor's office and state which one has the shorter waiting times.

Dr Watt		Stem		Dr Who		
	7 4	3	2	6	8	
	5 2	4	1	3	3	9
7 4	3	5	2	3	7	
8 2	2	6	1	4		

12-08 Interquartile range

20 Find the upper quartile, lower quartile and interquartile range for this data set.

   42   44   41   48   49   45   46   47   44   43

12-09 Box-and-whisker plots

21 This box-and-whisker plot shows the number of customers visiting a convenience store each day.

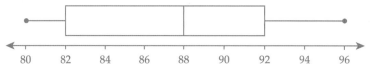

   80   82   84   86   88   90   92   94   96

  a   What was the range?

  b   Find the interquartile range.

  c   On what percentage of days were the number of customers over 82?

  d   What statistical name is given to the score of 88?

EQUATIONS AND INEQUALITIES 13

IN THIS CHAPTER YOU WILL:

- solve two-step equations
- solve equations with variables on both sides
- expand and solve equations with brackets
- solve simple quadratic equations
- use equations to solve practical problems
- solve inequalities and graph their solutions on a number line

Shutterstock.com/Krivosheev Vitaly

WORDBANK

equation A number sentence that contains algebraic terms, numbers and an equals sign; for example, $x - 3 = 8$.

solve an equation To find the value of the variable that makes the equation true.

inverse operation The 'opposite' process; for example, the inverse operation to adding (+) is subtracting (−).

To solve an equation:
- do the same to both sides of the equation: this will keep it balanced
- use inverse operations to simplify the equation (+ and − are inverse operations, × and ÷ are inverse operations)
- write the **solution** (answer) as: x = a number

EXAMPLE 1

Use inverse operations to solve each equation.

a $4x + 3 = 15$ **b** $2n - 8 = 10$ **c** $\dfrac{x - 2}{3} = 5$ **d** $24 - 3w = 36$

✱ These are called **two-step equations** because they require two steps to solve.

SOLUTION

a
$$4x + 3 = 15$$
$$4x + 3 - 3 = 15 - 3 \quad \longleftarrow \quad -3 \text{ from both sides}$$
$$4x = 12$$
$$\frac{4x}{4} = \frac{12}{4} \quad \longleftarrow \quad \div \text{ both sides by 4}$$
$$x = 3$$
(Check: $4 \times 3 + 3 = 15$)

b
$$2n - 8 = 10$$
$$2n - 8 + 8 = 10 + 8 \quad \longleftarrow \quad + 8 \text{ to both sides}$$
$$2n = 18$$
$$\frac{2n}{2} = \frac{18}{2} \quad \longleftarrow \quad \div \text{ both sides by 2}$$
$$n = 9$$
(Check: $2 \times 9 - 8 = 10$)

c
$$\frac{x - 2}{3} = 5$$
$$\left(\frac{x - 2}{3}\right) \times 3 = 5 \times 3 \quad \longleftarrow \quad \times \text{ both sides by 3}$$
$$x - 2 = 15$$
$$x - 2 + 2 = 15 + 2 \quad \longleftarrow \quad + 2 \text{ to both sides}$$
$$x = 17$$
(Check: $\dfrac{17 - 2}{3} = 5$)

d
$$24 - 3w = 36$$
$$24 - 3w - 24 = 36 - 24 \quad \longleftarrow \quad -24 \text{ from both sides}$$
$$-3w = 12$$
$$\frac{-3w}{-3} = \frac{12}{-3} \quad \longleftarrow \quad \div \text{ both sides by } (-3)$$
$$w = -4$$
(Check: $24 - 3 \times (-4) = 36$)

1   To solve $4x + 2 = 12$, which operation would you do first? Select the correct answer **A**, **B**, **C** or **D**.

   A $\div 4$ **B** -2 **C** $\times 4$ **D** $+2$

2   To solve $15 - 3n = -10$, which operation would you do first? Select **A**, **B**, **C** or **D**.

   A $+15$ **B** $\div (-3)$ **C** -15 **D** $+3$

3   **a** Make up a one-step equation that has $x = 3$ as its solution.

   b Make up a two-step equation that has $x = 3$ as its solution.

   c Make up another two-step equation that has $x = 3$ as its solution.

4   Copy and complete the solution to each equation.

   a $3a - 5 = 19$

   $3a - 5 + \underline{} = 19 + \underline{}$

   $3a = \underline{}$

   $\dfrac{3a}{} = \dfrac{24}{}$

   $a = \underline{}$

   b $5x + 4 = 24$

   $5x + 4 - \underline{} = 24 - \underline{}$

   $5x = \underline{}$

   $\dfrac{}{5} = \dfrac{20}{}$

   $x = \underline{}$

5   Solve each equation. Check your answers by substituting back into the original equation.

   a $a - 4 = 6$ **b** $w + 3 = 8$ **c** $4n = 72$ **d** $\dfrac{2x}{3} = 6$

   e $3x - 4 = 5$ **f** $4b - 1 = 23$ **g** $5 + 2c = 23$ **h** $10 - 3a = -11$

   i $2x - 1 = 6$ **j** $3x - 2 = 10$

6   Solve each equation. The solutions may be negative or fractions.

   a $3x + 5 = 12$ **b** $3n - 4 = 12$

   c $16 - 2a = -10$ **d** $9 + 3x = 15$

   e $18 - 4s = 21$ **f** $7y - 4 = -12$

   g $15 - 3r = -6$ **h** $11w - 3 = -31$

   i $\dfrac{x - 3}{4} = 5$ **j** $\dfrac{n + 5}{4} = -9$

   k $\dfrac{8 - u}{5} = 5$ **l** $\dfrac{-12 - x}{4} = 8$

7   Is each statement true or false?

   a $x = 3$ is the solution to $2x - 3 = 3$

   b $n = -1$ is the solution to $\dfrac{n - 4}{5} = 1$

   c $c = -2$ is the solution to $14 - 2c = 18$

   d $m = 5$ is the solution to $3m + 7 = 23$

To solve an equation with variables on both sides:
- use inverse operations to move all the variables to the left-hand side (LHS) of the equation
- use inverse operations to move all the numbers to the right-hand side (RHS) of the equation
- then solve the equation

EXAMPLE 2

Solve each equation.

a $2x - 5 = x + 14$

b $4n + 8 = 2n - 16$

SOLUTION

✳ Move all variables to the LHS of the equation.

a
$$2x - 5 = x + 14$$
$$2x - 5 - x = x + 14 - x \quad \longleftarrow \quad -x \text{ from both sides}$$
$$x - 5 = 14$$
$$x - 5 + 5 = 14 + 5 \quad \longleftarrow \quad +5 \text{ to both sides}$$
$$x = 19$$

b
$$4n + 8 = 2n - 16$$
$$4n + 8 - 2n = 2n - 16 - 2n \quad \longleftarrow -2n \text{ from both sides}$$
$$2n + 8 = -16$$
$$2n + 8 - 8 = -16 - 8 \quad \longleftarrow \quad -8 \text{ from both sides}$$
$$2n = -24$$
$$\frac{2n}{2} = \frac{-24}{2} \quad \longleftarrow \quad \div \text{ both sides by 2}$$
$$n = -12$$

Check that LHS = RHS:

LHS = $2x - 5$
$= 2 \times 19 - 5$
$= 33$
RHS = $x + 14$
$= 19 + 14$
$= 33$
LHS = RHS, so the solution $x = 19$ is correct.

LHS = $4n + 8$
$= 4 \times (-12) + 8$
$= -40$
RHS = $2n - 16$
$= 2 \times (-12) - 16$
$= -40$
LHS = RHS, so the solution $n = -12$ is correct.

Getty images/Dave King

1   To solve $3x - 6 = x + 12$, which operation would you do first? Select the correct answer **A, B, C** or **D**.

    A $+ x$ **B** $+ 12$ **C** $- x$ **D** $- 6$

2   To solve $5x + 11 = 3x - 7$, which operation would you do first? Select **A, B, C** or **D**.

    A $- 3x$ **B** $- 7$ **C** $+ 3x$ **D** $+ 11$

3   Write down the LHS of each equation.

    a $3x - 5 = 2x + 7$ **b** $4a + 9 = 2a - 5$ **c** $6 - 2x = x + 18$

4   Write down the RHS of each equation in Question **3**.

5   Solve each equation in Question **3** and check that each solution is correct.

6   Copy and complete the solution to each equation.

    a $3w + 18 = w - 4$ **b** $5x - 19 = 4x + 17$

          $3w + 18 - \_\_ = w - 4 - \_\_$ $5x - 19 - \_\_ = 4x + 17 - \_\_$

               $2w + \_\_ = -4$ $x - \_\_ = 17$

          $2w + 18 - \_\_ = -4 - \_\_$ $x - 19 + \_\_ = 17 + \_\_$

                     $w = \_\_\_$ $x = \_\_\_$

7   Solve each equation.

    a $5x - 8 = 12 + 3x$ **b** $3n + 5 = 11 + n$

    c $26 + 4x = 3x - 12$ **d** $6d - 8 = 4d + 10$

    e $4m - 3 = -15 + 3m$ **f** $6 - 8w = 9 - 5w$

    g $6 + 2x = 17 + x$ **h** $15 + 5n = 2n + 12$

    i $25 - 3x = 15 - 4x$

8   Is each statement true or false?

    a $x = 6$ is the solution to $2x - 3 = x + 3$

    b $x = -1$ is the solution to $4 - x = 2x + 1$

    c $c = -2$ is the solution to $14 - 2c = 20 + c$

    d $m = 5$ is the solution to $3m + 7 = 49 - m$

9   Solve each equation. Some answers may be negative or fractions.

    a $5x - 6 = 13 + 3x$ **b** $22 - 3n = 11 + 2n$

    c $26 + 4x = 3x - 5$ **d** $8d - 12 = 3d$

    e $20 - 4m = -15 + 2m$ **f** $-7w = 12 - 5w$

    g $6 - 2x = 8 + 3x$ **h** $28 - 5n = -2n$

    i $21 - 3x + 4 = 10 - 4x$

An equation with brackets, such as $3(x - 2) = 9$, can be solved by expanding out the brackets first and then solving as before using inverse operations.

EXAMPLE 3

Solve each equation.

a $2(x - 4) = 6$ **b** $3(2a + 5) = -9$ **c** $-3(4x - 2) = 10$

SOLUTION

a
$$2(x - 4) = 6$$
$$2x - 8 = 6 \quad \longleftarrow \text{Expand}$$
$$2x - 8 + 8 = 6 + 8 \quad \longleftarrow +8 \text{ to both sides}$$
$$2x = 14$$
$$\frac{2x}{2} = \frac{14}{2} \quad \longleftarrow \div \text{ both sides by 2}$$
$$x = 7$$

Check that LHS = RHS:
LHS $= 2 \times (7 - 4)$
$\quad = 2 \times 3$
$\quad = 6$
RHS $= 6$
LHS = RHS so the solution $x = 7$ is correct.

b
$$3(2a + 5) = -9$$
$$6a + 15 = -9 \quad \longleftarrow \text{Expand}$$
$$6a + 15 - 15 = -9 - 15 \quad \longleftarrow -15 \text{ from both sides}$$
$$6a = -24$$
$$\frac{6a}{6} = \frac{-24}{6} \quad \longleftarrow \div \text{ both sides by 6}$$
$$a = -4$$

LHS $= 3 \times [2 \times (-4) + 5]$
$\quad = 3 \times (-3)$
$\quad = -9$
RHS $= -9$
LHS = RHS so the solution $a = -4$ is correct.

c
$$-3(4x - 2) = 10$$
$$-12x + 6 = 10$$
$$-12x + 6 - 6 = 10 - 6$$
$$-12x = 4$$
$$\frac{-12x}{-12} = \frac{4}{-12}$$
$$x = -\frac{1}{3}$$

Check that LHS = RHS:
LHS $= -3 \times [4 \times \left(-\frac{1}{3}\right) - 2]$

$\quad = -3 \times \left(-\frac{10}{3}\right)$
$\quad = 10$
RHS $= 10$
LHS = RHS so the solution $x = -\frac{1}{3}$ is correct.

1 Find the solution to $2(x - 5) = 6$. Select the correct answer **A**, **B**, **C** or **D**.

    A $x = 4$ **B** $x = 8$ **C** $x = -8$ **D** $x = 6$

2 Find the solution to $-2(a + 6) = -8$. Select **A**, **B**, **C** or **D**.

    A $a = 4$ **B** $a = 2$ **C** $a = -2$ **D** $a = -4$

3 True or false?

    a $3(a + 2) = 3a + 2$ **b** $-4(w - 2) = -4w - 8$

    c $4(2x + 1) = 8x + 4$ **d** $-2(5n - 3) = -10n + 6$

4 Copy and complete the solution to each equation.

    a $2(w - 4) = 14$ **b** $-3(a + 5) = 12$

               $2w - \_\_\_ = 14$ $-3a - 15 = \_\_\_$

           $2w - 8 + \_\_\_ = 14 + \_\_\_$ $-3a - 15 + \_\_\_ = 12 + \_\_\_$

                  $2w = \_\_\_$ $-3a = \_\_\_$

                    $w = \_\_\_$ $a = \_\_\_$

5 Solve each equation and check your solutions.

    a $2(x - 6) = 18$ **b** $3(n + 4) = 24$

    c $4(x - 6) = 20$ **d** $8(d - 5) = 16$

    e $-2(m + 3) = -10$ **f** $-7(w - 3) = -14$

    g $6(2x - 1) = 18$ **h** $5(3n + 4) = 50$

    i $-3(2x + 4) = 6$

6 True or false?

    a $x = 6$ is the solution to $2(x - 3) = 8$

    b $x = -1$ is the solution to $4(2x - 1) = -12$

    c $c = -2$ is the solution to $5(c + 6) = 24$

    d $m = 8$ is the solution to $3(2m + 7) = 68$

7 Solve each equation. The solutions may be negative or fractions.

    a $5(x - 6) = 13$ **b** $3(n + 4) = 14$

    c $4(x - 3) = -20$ **d** $8(d - 9) = 36$

    e $-4(m - 5) = -15$ **f** $-7(2w + 3) = 12$

    g $-3(2x - 5) = 8$ **h** $-2(5n - 4) = 16$

    i $-8(3x + 4) = 60$

8 Write an equation with brackets that has a solution of $x = 3$.

9 Write an equation with brackets that has a solution of $x = -3$.

WORDBANK

quadratic equation An equation such as $x^2 = 4$, where the highest power of the variable is 2, that is, the variable squared.

surd A square root whose answer is not an exact decimal number; for example, $\sqrt{8} = 2.8284...$ is a surd because there isn't an exact number squared that is equal to 8.

To solve the quadratic equation $x^2 = 4$, we need to find the number that when squared is equal to 4.
$2^2 = 4$ but also $(-2)^2 = 4$ as $-2 \times (-2) = 4$.
So the solution to $x^2 = 4$ is $x = 2$ or $x = -2$, which we write as $x = \pm 2$.
Sometimes, the solution to a quadratic equation is a **surd**. For example, if $x^2 = 5$, then $x = \pm\sqrt{5}$, which cannot be simplified.

EXAMPLE 4

Solve each equation, writing each solution in exact form.

✳ | exact form means as a number written as a decimal or surd, without rounding.

a $x^2 = 16$ **b** $x^2 = 7$ **c** $2a^2 = 18$ **d** $3x^2 = 45$

SOLUTION

a $x^2 = 16$
$x = \pm\sqrt{16}$
$x = \pm 4$

b $x^2 = 7$
$x = \pm\sqrt{7}$

c $2a^2 = 18$
$\dfrac{2a^2}{2} = \dfrac{18}{2}$
$a^2 = 9$
$a = \pm\sqrt{9}$
$a = \pm 3$

d $3x^2 = 45$
$\dfrac{3x^2}{3} = \dfrac{45}{3}$
$x^2 = 15$
$x = \pm\sqrt{15}$

x^2 and $\sqrt{}$ are inverse operations.
$\pm\sqrt{7}$ and $\pm\sqrt{15}$ are surds.

EXAMPLE 5

Solve $4x^2 = 50$, writing the solution correct to 2 decimal places.

SOLUTION

$4x^2 = 50$
$\dfrac{4x^2}{4} = \dfrac{50}{4}$
$x^2 = 12.5$
$x = \pm\sqrt{12.5}$
$x = \pm 3.5355...$
$x \approx \pm 3.54$

1   What is the solution to $x^2 = 25$ in simplest form? Select the correct answer **A**, **B**, **C** or **D**.
    A $\pm\sqrt{25}$ **B** ± 5 **C** $\mp\sqrt{25}$ **D** ± 12.5

2   What is the solution to $x^2 = 6$ correct to two decimal places? Select **A**, **B**, **C** or **D**.
    A $\pm\sqrt{6}$ **B** ± 2.44 **C** ± 2.45 **D** ± 3

3   Is each statement true or false?
    a   If $x^2 = 36$, then $x = \pm 6$ **b**   If $x^2 = 3$, then $x = \pm\sqrt{13}$
    c   If $x^2 = 121$, then $x = \pm 11$ **d**   If $x^2 = 82$, then $x = \pm 9$

4   Solve each equation, leaving your answer in exact form.
    a $x^2 = 49$ **b** $x^2 = 64$
    c $a^2 = 144$ **d** $n^2 = 1$
    e $x^2 = 11$ **f** $x^2 = 19$
    g $w^2 = 24$ **h** $b^2 = 39$
    i $x^2 = -49$ **j** $x^2 = 65$
    k $a^2 = 1000$ **l** $n^2 = -12$

5   Write each solution to Question 4 correct to one decimal place.

6   Solve each equation, leaving your answer in exact form.
    a $2x^2 = 72$ **b** $3x^2 = 27$
    c $5a^2 = 125$ **d** $7n^2 = 7$
    e $4x^2 = 7$ **f** $8x^2 = 116$
    g $3w^2 = 159$ **h** $8b^2 = 32$
    i $5x^2 = 490$ **j** $6x^2 = -36$
    k $4a^2 = 1200$ **l** $8n^2 = 648$

7   Write each solution to Question 6 correct to 2 decimal places.

8   Is each statement true or false?
    a $x^2 = c$ has two solutions if c is a positive number.
    b $x^2 = c$ always has two solutions.
    c $x^2 = c$ can have three solutions.
    d $x^2 = c$ can have only one solution.
    e   The solutions to $x^2 = c$ are always whole numbers or fractions.
    f $x^2 = c$ has two solutions if c is a negative number.

9   Solve each equation correct to 2 decimal places.
    a $n^2 = 431$ **b** $2w^2 = 18.62$
    c $4b^2 = 21.6$ **d** $9u^2 = 9000$

10 Solve $5x^2 = 500\ 000\ 000$.

To solve a word problem involving equations:
- let the unknown value be x (or another variable)
- convert the words into an equation
- solve the equation
- answer the problem in words

EXAMPLE 6

Kane is trying to guess Natasha's favourite number. Natasha says if she triples the number and subtracts 7, the result is 26. What is the number?

SOLUTION

Let the number be x.

$3x - 7 = 26$ ⟵ Triple the number is 3 times the number.

$3x - 7 + 7 = 26 + 7$ ⟵ Solve the equation.

$3x = 33$

$\dfrac{3x}{3} = \dfrac{33}{3}$

$x = 11$

∴ Natasha's favourite number is 11.
(Check: $3 \times 11 - 7 = 26$)

EXAMPLE 7

James is twice Tayla's age. In 3 years' time, the sum of their ages will be 42.

How old are Tayla and James now?

SOLUTION

Let Tayla's age now be n. So James' age is $2n$.

In 3 years, Tayla will be $n + 3$ (3 years older) and James will be $2n + 3$ (also 3 years older).

In 3 years, the sum of their ages is 42.

$n + 3 + 2n + 3 = 42$

$3n + 6 = 42$ ⟵ Simplifying

$3n + 6 - 6 = 42 - 6$

$3n = 36$

$\dfrac{3n}{3} = \dfrac{36}{3}$

$n = 12$

∴ Tayla is 12 years old now and James is ($2 \times 12 =$) 24 years old.

(Check: In 3 years, the sum of their ages will be $15 + 27 = 42$)

1 Which equation says 'Twice a number n plus three is equal to fifteen'? Select the correct answer **A**, **B**, **C** or **D**.

   A $n + 3 = 15$ **B** $2n - 3 = 15$ **C** $2n + 15 = 3$ **D** $2n + 3 = 15$

2 Find the solution to the equation in Question **1**. Select **A**, **B**, **C** or **D**.

   A $n = 9$ **B** $n = 6$ **C** $n = 12$ **D** $n = -6$

3 Convert each worded statement into an equation, then solve the equation.

   a The sum of a number n and 5 is equal to 9.

   b The difference between a number x and 7 is 12.

   c The product of a number w and 6 is 48.

   d The quotient of a number m and 4 is 13.

4 Ryan doubles his lucky number and adds 3. The result is 17. What is Ryan's lucky number?

5 Laura thinks of a number and then subtracts five from the number. She then multiplies by four and adds 2. The result is 34. What is the number that Laura first thought of?

6 What is the number that, when added to seven and then multiplied by eight, is equal to 96?

7 Twice a number plus five is equal to three times the number less nine. What is the number?

8 Four times a number minus eight is equal to three times the number less two.

9 The perimeter of this square is 60 cm.
   What is the value of s?

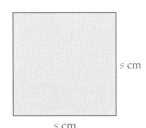

s cm

s cm

10 The area of a triangle is 68 cm². What is the height of the triangle if its base is 8 cm?

11 The perimeter of this rectangle is 114 cm.
   What is the value of b?

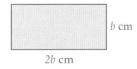

b cm

$2b$ cm

12 An electrician uses the formula $C = 45h + 80$ to calculate his charge. C is his charge in dollars and h is the number of hours he works on the job. Calculate the number of hours he worked if he charged $1275.

13 Katie is twice Amy's age. The sum of their ages is 54. How old are Katie and Amy?

14 Jasmine is triple Trent's age. In four years time, the sum of their ages will be 36. How old are Jasmine and Trent now?

15 The population of Mt Gotham is twice that of View Valley. If the sum of both populations is 270 000, what is the population of Mt Gotham?

WORDBANK

inequality A number sentence contains algebraic terms, numbers and an inequality sign; for example, $x + 9 > 15$, $4x - 2 \le 8$, where

< means 'is less than', > means 'is greater than'

≤ means 'is less than or equal to', ≥ means 'is greater than or equal to'

An equation like $4x - 2 = 8$ has one value of x as its solution, but an inequality like $4x - 2 \le 8$ has a **range of values of** x as its solution.

EXAMPLE 8

Solve each inequality and graph its solution on a number line.

a $x - 2 < 1$

b $5a + 4 \ge 19$

SOLUTION

Solve each inequality using inverse operations, as if it was an equation.

a
$$x - 2 < 1$$
$$x - 2 + 2 < 1 + 2 \quad \longleftarrow \quad \text{+ 2 to both sides}$$
$$x < 3$$

> ✱ All values less than 3, say –1, 2, 0, 2.9, are solutions to the inequality $x - 2 < 1$.

(Check using $x = 2$: $2 - 2 = 0 < 1$)

Graphing the solution:

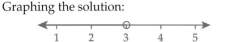

> ✱ The open circle on 3 means that 3 is not included in the solution $x < 3$.

b
$$5a + 4 \ge 19$$
$$5a + 4 - 4 \ge 19 - 4 \quad \longleftarrow \quad \text{– 4 from both sides}$$
$$5a \ge 15$$
$$\frac{5a}{5} \ge \frac{15}{5} \quad \longleftarrow \quad \div \text{ both sides by 5}$$
$$a \ge 3$$

(Check using $a = 4$: $5 \times 4 + 4 = 24 \ge 19$)

> ✱ The closed circle on 3 means that 3 is included in the solution $a \ge 3$.

To solve an inequality:
- ▨ do the same to both sides
- ▨ use inverse operations to simplify the inequality
- ▨ if **multiplying** or **dividing** both sides by a **negative** number, **reverse** the inequality sign

---

EXAMPLE 9

Solve each inequality and graph its solution on a number line.

a $2(5 - x) \le 20$

b $\dfrac{6 - 3a}{4} > 12$

13-06 | Inequalities

SOLUTION

a

$$2(5 - x) \le 20$$
$$10 - 2x \le 20$$
$$10 - 2x - 10 \le 20 - 10$$
$$-2x \le 10$$
$$\frac{-2x}{-2} \ge \frac{10}{-2}$$

✱ Dividing by (−2) requires changing ≤ to ≥

$$x \ge -5$$

b

$$\frac{6 - 3a}{4} \times 4 > 12 \times 4$$
$$6 - 3a > 48$$
$$6 - 3a - 6 > 48 - 6$$
$$-3a > 42$$
$$\frac{-3a}{-3} < \frac{42}{-3}$$
$$a < -14$$

EXERCISE 13–06

1 Solve $2x - 3 \le 5$. Select the correct answer **A, B, C** or **D**.

   A $x \ge 4$ **B** $x \le 1$ **C** $x \le 4$ **D** $x \ge 1$

2 Solve $8 - 3x \ge 14$. Select **A, B, C** or **D**.

   A $x \ge -2$ **B** $x \le -2$ **C** $x \le 2$ **D** $x \ge 2$

3 Copy and complete the solution to each equation.

   a $3x + 1 < 13$

       $3x + 1 - \underline{\quad} < 13 - \underline{\quad}$

           $3x < \underline{\quad}$

           $\dfrac{3x}{3} < \dfrac{12}{}$

           $x < \underline{\quad}$

   b $15 - 2x \ge 21$

       $15 - 2x - \underline{\quad} \ge 21 - \underline{\quad}$

           $-2x \underline{\quad} 6$

           $\dfrac{-2x}{} \le \dfrac{6}{}$

           $x \le \underline{\quad}$

4 Solve each inequality and graph its solution on a number line. Check your solutions by substituting back into the original inequality.

   a $2x - 4 < 12$ **b** $5a + 2 > 17$ **c** $2b + 5 \ge 23$

   d $4a - 8 \le 24$ **e** $6w - 5 > 25$ **f** $b - 11 > -14$

   g $3n + 1 < 49$ **h** $7m + 3 > 31$ **i** $9 - x \le 18$

   j $15 - 2a > 12$ **k** $8 - 4x < 32$ **l** $12 - 6b \ge 60$

5 Solve each inequality.

   a $2(x - 4) < 16$ **b** $5(2 - v) < 35$ **c** $2(3x + 2) > 76$

   d $3(4 - 5a) < 72$ **e** $\dfrac{2x - 1}{3} \le 5$ **f** $\dfrac{3a + 2}{4} > 8$

   g $\dfrac{6 - 3n}{2} \ge 12$ **h** $\dfrac{8 - 3b}{3} < 7$ **i** $\dfrac{2(3x - 5)}{3} \le 8$

6 Make up an inequality that has a solution of:

   a $x > 1$ **b** $x < 4$ **c** $x \ge -2$ **d** $x \le 9$

CODE PUZZLE

Match each equation on the left with its solution on the right.

1	$3x - 2 = 19$		A	-2
2	$8 - 3x = 32$		B	-3
3	$2x + 5 = -1$		C	4
4	$9 + 4x = 21$		D	-8
5	$2(7 - x) = 18$		E	18
6	$3(2x + 1) = 57$		H	24
7	$-2(4 - 2x) = 8$		L	7
8	$7(3x - 5) = 91$		N	-24
9	$\dfrac{2x + 3}{3} = 13$		O	3
10	$\dfrac{x - 5}{2} = 16$		R	6
11	$\dfrac{x}{3} + 4 = 12$		T	9
12	$8 - \dfrac{x}{2} = 20$		Y	37

Use the matched question numbers and answer letters to decode the answer to the riddle below.

What do a healthy diet and an equation have in common?

6–11–9–10   5–8–9   3–4–6–11   3–5–1–5–12–7–9–2

Part A General topics

Calculators are not allowed.

1 Solve $3(x + 10) = 12$

2 Evaluate 5.728×100.

3 Write 10:38 p.m. in 24-hour time.

4 How many hours are there between 2 a.m. and 7 p.m.?

5 Find x, giving a reason.

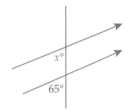

6 Copy and complete: $63 : 72 = 7 :$ ____

7 This cube has an open top. How many faces does it have?

6 m

8 What is the surface area of the cube in Question 7?

9 Convert $\dfrac{40}{6}$ to a mixed numeral.

10 What is northwest as a three-figure bearing?

Part B Equations and inequalities

Calculators are allowed.

13–01 Two-step equations

11 Solve $2n + 5 = 29$. Select the correct answer **A, B, C** or **D**.

   A $n = 12$ **B** $n = 17$ **C** $n = 24$ **D** $n = -12$

12 Solve $8 - 3x = 14$. Select **A, B, C** or **D**.

   A $n = 7$ **B** $n = 2$ **C** $n = -2$ **D** $n = -3$

13–02 Equations with variables on both sides

13 Solve each equation.

   a $3a + 15 = 2a - 4$

   b $5n + 8 = 3n - 6$

13–03 Equations with brackets

14 Solve each equation.

   a $-2(a - 7) = 24$

   b $9(2x - 3) = -27$

15 Make up an equation that has a solution of $x = -3$.

13–04 Simple quadratic equations $ax^2 = c$

16 Solve each equation, leaving the answers in exact form.

   a $4x^2 = 64$

   b $3n^2 = 39$

17 Solve $2y^2 = 10$, writing the solutions correct to 2 decimal places.

13–05 Equation problems

18 Jackson is three times Amanda's age. In 4 years time, the sum of their ages will be 84. How old are Jackson and Amanda now?

13–06 Inequalities

19 Solve each inequality and graph the solution on a number line.

   a $5x - 3 < 27$

   b $-2a + 30 \leq 24$

PROBABILITY

14

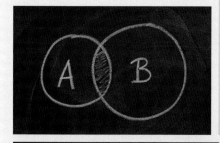

IN THIS CHAPTER YOU WILL:

- calculate the probability of simple events
- calculate the probability of complementary events
- find the relative frequency of an event
- interpret and draw Venn diagrams and use them to solve probability problems
- read two-way tables and use them to solve probability problems

Shutterstock.com/marekuliasz

WORDBANK

probability The chance of an event occurring, measured on a scale from 0 to 1.

random Where every possible outcome is equally likely to occur.

event An outcome or group of outcomes.

sample space The set of all possible outcomes in a situation.

complementary event The 'opposite' event, or the event not taking place. For example, the complement of rolling 5 on a die is rolling 1, 2, 3, 4 or 6.

If all possible outcomes are **equally likely**, then the **probability of an event**, $P(E)$, is:

$$P(E) = \frac{\text{number of outcomes in the event}}{\text{number of outcomes in the sample space}}$$

The complement of E is written as $\bar{E}$.

- $P(E) + P(\bar{E}) = 1$
- or $P(\bar{E}) = 1 - P(E)$
- or $P(\text{not } E) = 1 - P(E)$.

EXAMPLE 1

A basket contains 6 red, 8 blue and 4 yellow socks. If a sock is selected at random from the bag, find the probability that it is:

a blue

b not blue

SOLUTION

a Total socks = 6 + 8 + 4

= 18

$$P(\text{blue}) = \frac{8}{18}$$

$$= \frac{4}{9}$$

b $P(\text{not blue}) = 1 - P(\text{blue})$

$$= 1 - \frac{4}{9}$$

$$= \frac{5}{9}$$

 This is also true because there are 10 socks out of 18 that are not blue, and $\frac{10}{18} = \frac{5}{9}$

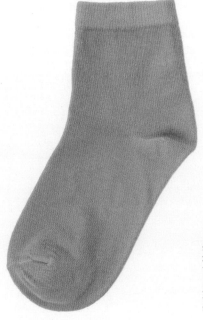

iStockphoto.com/puleka19

1   If the probability that it will rain tomorrow is $\dfrac{3}{5}$, what is the probability that it will not rain tomorrow? Select the correct answer **A**, **B**, **C** or **D**.

   A $\dfrac{2}{5}$ **B** $\dfrac{3}{5}$ **C** 1   **D** $\dfrac{4}{5}$

2   List the sample space (the set of all possible outcomes) for each chance experiment.
   a   Tossing a coin.
   b   Selecting an odd number between 22 and 36.
   c   Tossing two coins.

3   When rolling a die, find the probability that the number that comes up is:
   a   6   b   3
   c   less than 4   d   greater than 3
   e   even   f   odd
   g   prime   h   a factor of 6

4   a   A box contains 5 black and 6 white chocolates. If one is chosen at random, what is the probability that it is:
      i   black?   ii   white?
   b   Add your probabilities in **i** and **ii** of part **a**. What is the sum of the probabilities?

5   Describe in words the complementary event for each event.
   a   Rolling 4 on a die
   c   Coming first in a race
   c   Choosing a yellow sock from a drawer containing red and yellow socks.
   d   Choosing a diamond card from a deck of cards
   e   Winning a soccer match.

6   A bag contains 3 blue, 4 red and 5 white marbles. What is the probability that a marble selected at random from the bag is:
   a   blue?   b   red?   c   white?
   d   not blue?   e   not red?   f   not white?

7   The probability that Chloe wins first prize in a raffle is $\dfrac{1}{1000}$.

   a   What is the probability that Chloe does not win first prize?
   b   If Chloe had bought 2 tickets, how many tickets were sold altogether?

8   One card is selected at random from a deck of playing cards. What is the probability that it is:
   a   a club?   b   a heart?   c   not a heart?
   d   a 7?   e   not a 7?   f   a Queen?

WORDBANK

frequency The number of times an outcome occurs.

relative frequency The frequency of an event compared to the total number of trials in an experiment, used to estimate the probability of the event.

$$P(E) = \frac{\text{number of times the event occurred}}{\text{total number of trials}}$$

$$= \frac{\text{frequency of the event}}{\text{total frequency}}$$

EXAMPLE 2

Two coins are tossed 40 times and the numbers of tails each time were recorded.

Number of tails	Frequency
0	15
1	20
2	5

a   Find the relative frequency of tossing 1 tail.

b   Predict how many times 2 tails will occur if these 2 coins are tossed 400 times.

c   What is the relative frequency of tossing 2 *heads*?

SOLUTION

a   Relative frequency of 1 tail $= \dfrac{20}{40}$ ← $\dfrac{\text{Frequency of the event}}{\text{Total frequency}}$

$= \dfrac{1}{2}$

b   Experimental $P(2 \text{ tails}) = \dfrac{5}{40} = \dfrac{1}{8}$

Expected number of '2 tails' $= \dfrac{1}{8} \times 400$

$= 50$

c   Experimental $P(2 \text{ heads}) = \text{Experimental } P(0 \text{ tails})$

$= \dfrac{15}{40}$

$= \dfrac{3}{8}$

1 If a coin is tossed 20 times and 12 heads are thrown, what is the relative frequency of throwing a tail?

2 If a die is rolled 40 times and a 3 comes up 8 times, what is the relative frequency of not rolling a 3? Select the correct answer **A**, **B**, **C** or **D**.

   A $\dfrac{3}{40}$ **B** $\dfrac{8}{40}$ **C** 1   **D** $\dfrac{32}{40}$

3 A die is rolled 180 times and the results are recorded below.

Score	Frequency	Relative frequency	Relative frequency (%)
1	28		
2	32		
3	27		
4	35		
5	31		
6	27		
Total	180		

   a Copy and complete the table. In the last column, express the relative frequency as a percentage correct to one decimal place.

   b Find the experimental probability of rolling a 4 on a die.

   c Find the number of 4s that you would expect when a die is tossed 600 times.

   d What is the relative frequency of rolling a 5?

   e What is the relative frequency of not rolling a 5?

4 In groups of 2 to 3, toss two coins 50 times and record the number of tails in a table similar to the one in Example **2**.

   a Find the relative frequency of tossing no tails.

   b Predict the frequency of tossing no tails when the coins are tossed 500 times.

   c Find the sum of the relative frequencies.

   d Find the number of times that you would expect no tails to occur if you tossed the coins 500 times.

5 Three coins are tossed 80 times and the results are recorded below.

Number of heads	Frequency	Relative frequency	Relative frequency (%)
0	9		
1	32		
2	28		
3	11		
Total	80		

   a Copy and complete the table.

   b Find the experimental probability of tossing more heads than tails.

   c If the coins were tossed 600 times, how many times would you expect to get all heads or all tails?

A **Venn diagram** uses circles to group items into categories. Most Venn diagrams involve circles that **overlap**. For example, A could represent runners and B could represent soccer players.

Then the shaded region would represent **A and B**, which means people who are both runners and soccer players.

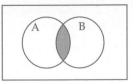

A or B means **A or B or both**

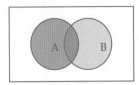

Runners or soccer players or both

A only means **A but not B**.

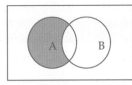

Runners only, not soccer players

A Venn diagram may involve two groups that do not overlap.

These categories are **mutually exclusive** as it is not possible to be both a boy and a girl.

EXAMPLE 3

25 students were surveyed on whether they liked seafood or chicken. The results are shown in the Venn diagram.

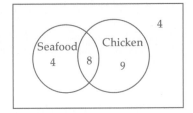

a How many students liked seafood only?

b How many students liked seafood or chicken?

c How many students liked seafood and chicken?

d What is the probability of choosing a student who likes chicken only?

e What is the probability that a student likes seafood or chicken but not both?

f What is the probability that a student likes neither seafood nor chicken?

SOLUTION

a 4 students liked seafood only.   ◄—— Seafood but not chicken.

b 21 students liked seafood or chicken.   ◄—— 4 + 8 + 9 = 21, seafood or chicken or both.

c 8 students liked seafood and chicken.   ◄—— The overlap section.

d $\dfrac{9}{25}$ ◄—— 9 students liked chicken, but not seafood.

e $\dfrac{13}{25}$ ◄—— 4 + 9 = 13 students like seafood or chicken, but not both.

f $\dfrac{4}{25}$ ◄—— 4 is the number outside the circles.

EXAMPLE 4

In 75 homes surveyed, 55 had dogs as pets, 32 had cats and 5 had neither dogs nor cats.
Show this information on a Venn diagram.

SOLUTION

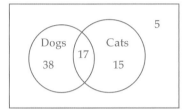

✱ **Aim to fill in the overlap section first.**

- 5 homes have neither dogs nor cats, so write 5 outside the circles
- This leaves 75 – 5 = 70 homes
- But 55 dogs + 32 cats = 87 pets, so the number of homes with both dogs and cats is 87 – 70 = 17
- So write 17 in the overlap
- This leaves 55 – 17 = 38 homes with dogs only
- This leaves 32 – 17 = 15 homes with cats only

✱ **Check that the numbers add up to 75.**

EXERCISE 14-03

1 40 students were surveyed on whether they liked basketball or netball. 32 students liked
   basketball, 19 liked netball and no-one said they didn't like both. How many students
   liked both sports? Select the correct answer **A**, **B**, **C** or **D**.

   A 10   **B** 11   **C** 12   **D** 13

2 In Question **1**, how many students liked netball but not basketball? Select **A**, **B**, **C** or **D**.

   A 8   **B** 11   **C** 19   **D** 21

3 The Venn diagram below shows how students travel to school each day.

   a How many students:

       i   were surveyed?

       ii   came by bus only?

       iii caught a train and bus?

       iv caught a train?

       v   caught neither a bus nor a train?

       vi did not catch a bus?

   b Are catching a train and catching a bus mutually exclusive?

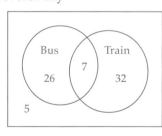

4   This Venn diagram shows the number of students playing video games.

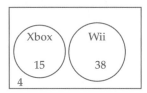

a   How many students:
   i   were surveyed?
   ii   play on the Wii?
   iii   play on the Xbox?
   iv   play on both?
   v   play on neither the Xbox nor Wii?
   vi   play on either the Xbox or Wii?
b   Is playing on the Xbox and the Wii mutually exclusive for this class of students?

5   In a group of 40 athletes, 23 are high jumpers, 26 are long jumpers and 6 are neither. Show this data on a Venn diagram.
a   How many athletes were:
   i   both high jumpers and long jumpers?
   ii   high jumpers but not long jumpers?
   iii   high jumpers or long jumpers but not both?
b   What is the probability of selecting an athlete who:
   i   is neither a high jumper nor long jumper?
   ii   is a long jumper?

6   The Venn diagram shows the types of food liked by a group of Year 10 students.
a   How many students:
   i   were surveyed?
   ii   liked Asian food only?
   iii   liked only one type of food?
b   What is the probability that a Year 10 student:
   i   liked only Italian?
   ii   liked neither Asian nor Italian?
   iii   liked Asian or Italian but not both?
   iv   liked at least one type of food?

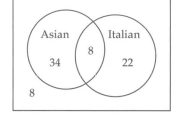

7   In a group of sports fans, 32 like watching football, 24 like cricket and 5 like neither. Of these, 7 like both football and cricket. Draw a Venn diagram to represent this information.
a   How many sports fans were surveyed?
b   What is the probability that a randomly-chosen sports fan likes:
   i   only cricket?
   ii   both sports?
   iii   only one of the sports?
   iv   football?

| Two-way tables

EXAMPLE 5

Sixty people were surveyed on whether they preferred city or country life.

The results are shown in the two-way table below.

	Male	Female	Total
City	16	22	38
Country	12	10	22
Total	28	32	60

a   How many people prefer the country?

b   How many males like the city?

c   What is the probability that a person selected at random will be female and like the country?

d   What is the probability of selecting a male?

SOLUTION

a   Total who preferred the country = 22

b   Number of males who prefer the city = 16

c $P(\text{female who likes the country}) = \dfrac{10}{60} = \dfrac{1}{6}$

d $P(\text{male}) = \dfrac{28}{60} = \dfrac{7}{15}$

EXAMPLE 6

This Venn diagram shows whether 64 students preferred yoghurt or ice-cream. Show this information on a two-way table.

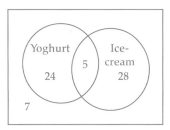

SOLUTION

✴ Start at the centre overlap section first.

- 5 students like both yoghurt and ice-cream
- 24 students like yoghurt, not ice-cream
- 28 students like ice-cream, not yoghurt
- 7 students like neither

Complete the totals column and row.

	Yoghurt	Not yoghurt	Total
Ice-cream	5	28	33
Not ice-cream	24	7	31
Total	29	35	64

1   If there are 22 students playing the flute and 26 students playing the violin and 42 students altogether, how many students play both instruments? Select the correct answer **A**, **B**, **C** or **D**.

   A 6   **B** 16   **C** 10   **D** 48

2   How many students play the flute only? Select **A**, **B**, **C** or **D**.

   A 18   **B** 20   **C** 22   **D** 16

redsnapper / Alamy

3   Copy and complete this two-way table.

	Female	Male	Total
Blonde hair	12	7	
Not blonde hair	26	19	
Total			

   a   How many people were surveyed?
   b   How many people had blonde hair?
   c   How many males did not have blonde hair?
   d   What is the probability that a person selected at random will be a blonde female?
   e   What is the probability of selecting a male?

4   Copy and complete this table showing girls and boys who like or don't like dancing.

	Girls	Boys	Total
Like dancing	18	14	
Don't like dancing	6	8	
Total			

   a   How many people liked dancing?
   b   How many girls did not like dancing?
   c   What is the probability that a student selected at random will be a boy who does not like dancing?
   d   What is the probability of selecting a girl who likes dancing?

5 a The Venn diagram shows whether 60 parents preferred TV shows or movies. Copy and complete the two-way table.

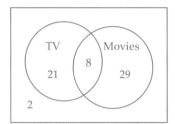

	TV	Not TV	Total
Movies			
Not movies			
Total			

b How many parents liked movies?

c How many parents liked TV but not movies?

d What is the probability that a parent selected at random will like both TV and movies?

e What is the probability of selecting a parent who likes TV or movies but not both?

iStockphoto.com

6 Copy and complete this table.

	Tennis	Not tennis	Total
Soccer		22	58
Not soccer	18		
Total		36	

a How many people played tennis?

b How many people played soccer but not tennis?

c What is the probability that a person selected at random plays tennis but not soccer?

d What is the probability of selecting a person who plays both sports?

EXAMPLE 7

Dylan selected a card at random from a deck of cards, recorded its suit and then returned it to the deck. He did this 100 times and the results were hearts (♥) 26, diamonds (♦) 22, spades (♠) 20 and clubs (♣) 32.

a Find, as a decimal, the experimental and theoretical probabilities for selecting:

   i a spades card   **ii** a black card

b Explain why the experimental and theoretical probabilities are different.

c What should happen to the experimental and theoretical probabilities if Dylan repeated his experiment 500 times instead of 100 times?

SOLUTION

a **i** Experimental P(spades card) $= \dfrac{20}{100} = 0.2$

       Theoretical P(spades card) $= \dfrac{13}{52} = \dfrac{1}{4} = 0.25$ ⟵   52 cards in a deck, 13 are spades

  ii Experimental P(black) $= \dfrac{32+20}{100} = \dfrac{52}{100} = 0.52$

       Theoretical P(black) $= \dfrac{26}{52} = 0.5$

b The experimental and theoretical probabilities are different as the experimental one is a result from an experiment and the theoretical is the actual probability.

c With more trials, the experimental probability should get closer to the theoretical probability.

EXAMPLE 8

A box of chocolates contains 6 strawberry, 5 mint and 7 caramel flavoured chocolates.

a If Amanda selected one from the box without looking, what is the probability that it was a mint chocolate?

b Amanda ate her chocolate and gave the box to her friend Amy. What is the probability that Amy randomly chooses a strawberry chocolate?

SOLUTION

a Total number of chocolates = 6 + 5 + 7 = 18

    P(mint) $= \dfrac{5}{18}$ ⟵   5 mint chocolates out of 18

b P(strawberry) $= \dfrac{6}{17}$ ⟵   6 strawberry chocolates out of 17 (1 chocolate removed)

1 If there are 9 red, 6 blue and 5 green marbles in a bag, what is the probability of choosing a green marble? Select the correct answer **A, B, C** or **D**.

   A $\dfrac{1}{4}$ **B** $\dfrac{6}{20}$ **C** $\dfrac{9}{20}$ **D** $\dfrac{3}{10}$

2 For the marbles in Question **1**, what is the probability of choosing a marble that is not blue? Select **A, B, C** or **D**.

   A $\dfrac{6}{20}$ **B** $\dfrac{7}{10}$ **C** $\dfrac{1}{4}$ **D** $\dfrac{3}{4}$

3 Lee chooses a letter of the alphabet from A to Z at random. What is the probability that this letter is:

   a a vowel?   **b** a consonant?   **c** from the word FUN?

   d from the word CARNIVAL?   **e** from the word SUCCESS?

4 A box of chocolates contains these flavours: 8 toffee, 4 cherry, 6 peppermint and 12 nutty. If Rhianna takes one chocolate at random, what is the probability that its flavour is

   a cherry?   **b** peppermint?   **c** nutty?   **d** not toffee?

5 Monique selected a card at random from a deck of cards, recorded its suit and then returned it to the pack. The results were hearts 54, diamonds 48, spades 42 and clubs 56.

   a Find the experimental probability of selecting a diamonds card:

      i as a fraction   **ii** as a decimal

   b Find the theoretical probability of selecting a diamonds card:

      i as a fraction   **ii** as a decimal

   c Are the experimental and theoretical probabilities from parts **a** and **b** similar?

   d What should happen to the experimental probability if Monique selects a card 500 times?

6 A jar of biscuits contains 10 chocolate biscuits, 7 shortbreads and 9 wafers. Zak chose one at random.

   a What is the probability that he chose:

      i a chocolate biscuit?

      ii a wafer or a shortbread biscuit?

      iii a biscuit that is not chocolate?

   b If Zak ate two of the wafers, what is the probability that he selects a shortbread biscuit next?

7 Andrea ordered two Super Supreme pizzas, one Hawaiian pizza and three Pepperoni pizzas.

   a If each pizza was cut into 8 equal pieces and Andrea chose one piece of pizza at random without looking, what is the probability that it is:

      i Super Supreme?   **ii** Pepperoni?

      iii Hawaiian or Pepperoni?   **iv** Not Super Supreme?

   b If Andrea was hungry and ate two pieces of Pepperoni, what is the chance that she will select a Super Supreme next?

FIND-A-WORD PUZZLE

Make a copy of this puzzle, then find all the words listed below in this grid of letters.

S	I	M	U	L	A	T	I	O	N	T	F	A	P
A	S	R	E	T	I	C	H	U	O	N	R	I	T
M	O	U	M	N	A	H	N	T	I	E	E	S	H
P	I	N	S	I	T	A	E	C	T	M	Q	L	Y
L	O	L	R	A	O	N	V	O	C	E	U	I	T
E	X	I	R	T	S	C	E	M	I	L	E	K	I
S	I	K	O	R	A	E	S	E	D	P	N	E	L
P	R	E	V	E	N	T	E	S	E	M	C	L	I
A	B	L	I	C	R	A	D	T	R	O	Y	Y	B
C	O	Y	L	B	A	B	O	R	P	C	R	I	A
E	X	P	E	R	I	M	E	N	T	A	L	A	B
P	Y	L	A	C	I	T	E	R	O	E	H	T	O
T	R	A	I	N	D	O	R	M	O	D	N	A	R
I	M	P	O	S	S	I	B	L	E	W	E	R	P

CERTAIN	CHANCE	COMPLEMENT	EVEN
EVENT	EXPERIMENTAL	FREQUENCY	IMPOSSIBLE
LIKELY	OUTCOMES	PREDICTION	PROBABILITY
PROBABLY	RANDOM	SAMPLE SPACE	SIMULATION
THEORETICAL	UNLIKELY		

PRACTICE TEST 14

Part A General topics

Calculators are not allowed.

1 Convert $\frac{3}{4}$ to a decimal.

2 Evaluate $4x^0 - (4x)^0$.

3 Write 495 000 in scientific notation.

4 How many hours and minutes are there between 4:30 a.m. and 7:45 p.m.?

5 What do x and y stand for in the formula for the area of a rhombus $A = \frac{1}{2}xy$?

6 Increase $70 by 20%.

7 Solve $5a - 6 = 3a + 20$.

8 Find the value of x.

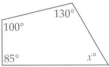

9 Find the mode of this data set.

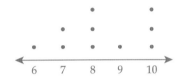

10 Find the range of the data in Question 9.

Part B Probability

Calculators are allowed.

14–01 Probability

11 A card is selected at random from a deck of playing cards. What is the probability of choosing a heart? Select the correct answer **A**, **B**, **C** or **D**.

   A $\frac{1}{2}$ **B** $\frac{1}{4}$ **C** $\frac{1}{13}$ **D** $\frac{1}{12}$

12 What is the complementary event to 'winning 3rd prize in a raffle'? Select **A**, **B**, **C** or **D**.

   A winning 1st prize   **B** not winning a prize

   C winning 2nd prize   **D** not winning 3rd prize

14–02 Relative frequency

13 Two coins were tossed 40 times. The results are recorded below.

Number of heads	Frequency	Relative frequency
0	6	
1	26	
2	8	
Total	40	

   a Copy and complete the table.

   b What is the probability of tossing one head?

   c If the coins were tossed 100 times, how many times would you expect to toss one head?

   d What is the relative frequency of tossing 2 *tails*?

14–03 Venn diagrams

14 Draw a Venn diagram illustrating the information below for 120 farms.

Farms with wheat crops   58

Farms with fruit trees   64

Farms with wheat crops and fruit trees   12

15 The Venn diagram shows how many students have smartphones and tablets.

a How many students were surveyed?

b How many students had tablets only?

c How many students had only one of the devices?

d What is the probability that a student has a smartphone but not a tablet?

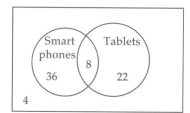

14–04 Two-way tables

16 Copy this table and complete it using the information from Question **15**.

	Smartphone	No smartphone	Total
Tablet			
No tablet			
Total			

17 Seventy people were asked whether they had a passport, with the results shown in the two-way table. Copy and complete the table.

	Male	Female	Total
Passport	32	25	
No passport	5	8	
Total			

a How many people had a passport?

b How many males did not have a passport?

c What is the probability that a person selected at random will be female with a passport?

14–05 Probability problems

18 The ten letters of the alphabet from A to J are written on separate pieces of paper and shuffled. Nina chooses one of these at random. What is the probability that this letter is:

a a vowel?   b a consonant?   c from the word PROBABILITY?

19 Emily chose a lolly at random from a jar containing 12 Minties, 6 chocolates and 14 Fantales.

a What is the probability that she chose:

     i a chocolate?   ii a lolly that was not a Fantale?

b If Emily ate two Fantales, what is the probability that she selects a Mintie next?

COORDINATE GEOMETRY

15

WHAT'S IN CHAPTER 15?

IN THIS CHAPTER YOU WILL:

- use a linear equation to complete a table of values
- find the rule (linear equation) for a table of values
- identify points and quadrants on a number plane
- graph tables of values on the number plane
- find the length and midpoint of an interval joining two points on the number plane
- find the gradient of an interval or line on a number plane

Shutterstock.com/ Policas

A **table of values** is a set of numbers listed in a table that follow an equation or formula relating two variables, say x and y. The values of y depend on the value of x.

> **To complete a table of values**, substitute each x-value into the equation to find the y-value.

EXAMPLE 1

Complete each table of values using the equation given.

a $y = x - 2$

x	0	1	2	3
y				

b $y = 3x + 1$

x	-1	1	2
y			

c $q = 4p - 2$

p	-2	0	3
q			

SOLUTION

Substitute the x-values from the table into each equation.

a $y = x - 2$
When $x = 0$, $y = 0 - 2 = -2$
When $x = 1$, $y = 1 - 2 = -1$, and so on.

x	0	1	2	3
y	-2	-1	0	1

b $y = 3x + 1$
When $x = -1$, $y = 3 \times (-1) + 1 = -2$
When $x = 1$, $y = 3 \times 1 + 1 = 4$, and so on.

p	-1	1	2
q	-2	4	7

c $q = 4p - 2$
When $p = -2$, $q = 4 \times (-2) - 2 = -10$
When $p = 0$, $q = 4 \times 0 - 2 = -2$, and so on.

p	-2	0	3
q	-10	-2	10

iStockphoto.com/FooTToo

1   Find the value of y when $x = 3$ if $y = 5x - 1$. Select the correct answer **A, B, C** or **D**.

   A 6   **B** 14   **C** 16   **D** 7

2   Find the value of m when $n = -4$ if $m = 12 - 2n$. Select **A, B, C** or **D**.

   A 10   **B** 14   **C** 4   **D** 20

3   If $y = 3x - 2$, then is each statement true or false?

   a When $x = 1$, $y = 1$ **b** When $x = -1$, $y = 0$

   c When $x = 2$, $y = -4$ **d** When $x = -2$, $y = -8$

   e When $x = 3$, $y = 4$ **f** When $x = -3$, $y = -11$

4   For $y = 14 - 3x$, find the value of y when:

   a $x = 1$ **b** $x = -1$ **c** $x = 2$

   d $x = -2$ **e** $x = 3$ **f** $x = -3$

5   Copy and complete each table.

   a $y = 3x$

x	-1	0	1	2
y				

   b $y = x + 4$

x	-1	0	1	2
y				

   c $y = 2x - 1$

x	-1	0	3	4
y				

   d $y = 3x + 2$

x	-2	0	1	3
y				

   e $y = 8 - x$

x	-2	0	2	4
y				

   f $y = 6 - 3x$

x	-3	0	3	6
y				

   g $d = 2c + 9$

c	-1	0	1	2
d				

   h $v = 4u - 7$

u	-1	0	1	2
v				

   i $t = 14 - 2s$

s	-4	-2	0	2
t				

   j $q = 7p - 5$

p	-2	0	2	4
q				

ISBN 9780170351058

We can find the rule (or equation or formula) of a table of values by looking for number patterns in the table.

EXAMPLE 2

Find the rule for each table of values.

a

x	−1	0	1	2
y	−2	0	2	4

b

p	2	3	4	5
q	−2	−1	0	1

SOLUTION

Look for a pattern between the x- and y-values and consider the four operations +, −, × and ÷.

a In each case, the x-value has been multiplied by 2 (× 2) to get the y-value: $-1 \times 2 = -2$, $0 \times 2 = 0$, and so on.
So the rule is: $y = x \times 2$, which can be simplified to $y = 2x$.

b In each case, the p-value has been reduced by 4 (− 4) to get the y-value: $2 - 4 = -2$, $3 - 4 = -1$, and so on.
So the rule is: $q = p - 4$.
Sometimes the rule is harder to find. In this case, try trial and error or use the method below.

> If the values in the top row are **consecutive** (increase by 1), look for a pattern in the values in the bottom row. This will help you to find the rule for the table of values.

EXAMPLE 3

Find the equation for each table of values.

a

x	−1	0	1	2
y	-5	-2	1	4

b

a	−1	0	1	2
b	6	4	2	0

SOLUTION

The rules for these tables of values are not so easy. They involve **two** operations: × and either + or −. Note that the values in the top row are consecutive (increase by 1).

Then look for a pattern in the values in the bottom row. This will give the **multiplier** number.

a

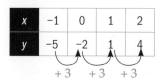

+3   +3   +3

y increases by 3 each time, so the multiplier is 3.

So the rule starts like this: $y = 3x$ _____, where we have to decide what goes in the blank.

Using the columns in the table:
$3 \times (-1) - 2 = -5$
$3 \times 0 - 2 = -2$
$3 \times 1 - 2 = 1$, and so on.
So the equation is: $y = 3x - 2$.

b

a	−1	0	1	2
b	6	4	2	0

$-2 \quad -2 \quad -2$

y decreases by 2 each time, so the multiplier is −2.

So the rule starts like this: $b = -2a$ ____.

Using the columns in the table:
$-2 \times (-1) + 4 = 6$
$-2 \times 0 + 4 = 4$, and so on.
So the equation is: $b = -2a + 4$.

EXERCISE 15-02

1 Find the rule connecting x and y if $y = 6$ when $x = 2$ and $y = 9$ when $x = 3$. Select the correct rule **A, B, C** or **D**.

  A $y = 2x$ **B** $y = -3x$ **C** $y = \dfrac{x}{3}$ **D** $y = 3x$

2 If $y = -8$ when $x = 2$ and $y = -12$ when $x = 3$, select the correct rule **A, B, C** or **D**.

  A $y = 2x$ **B** $y = -4x$ **C** $y = \dfrac{x}{4}$ **D** $y = 4x$

3 Match each table of values to its rule below.

a

x	−1	0	1	2
y	2	3	4	5

b

x	−2	0	2	6
y	−1	0	1	3

c

x	−1	0	1	2
y	6	5	4	3

d

x	−1	0	1	2
y	−3	0	3	6

e

x	−1	0	1	2
y	−5	−4	−3	−2

  A $y = x - 4$ **B** $y = 3x$ **C** $y = x + 3$ **D** $y = \dfrac{x}{2}$ **E** $y = 5 - x$

4 Find the equation for each table of values.

a

x	−1	0	1	2
y	3	4	5	6

b

x	−1	0	1	2
y	−7	−6	−5	−4

c

x	−1	0	1	2
y	3	5	7	9

d

x	−1	0	1	2
y	5	3	1	−1

e

x	−1	0	1	2
y	−2	1	4	7

f

x	−1	0	1	2
y	4	2	0	−2

A **number plane** is a grid for plotting points and drawing graphs.
It has an **x-axis**, which is horizontal (goes across), and a **y-axis**, which is vertical (goes up and down).
The **origin** is the centre of the number plane.

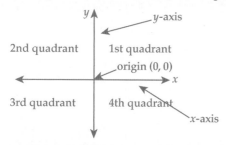

The number plane is divided into 4 quadrants (quarters).

EXAMPLE 4

Plot each point on a number plane.

A (1, 3)	B (–1, 3)	C (–2, –4)	D (3, –4)
E (0, 2)	F (–3, 0)	G (3, 0)	H (0, –2)

SOLUTION

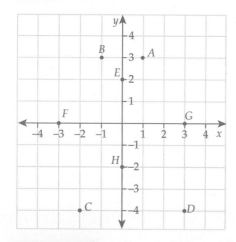

$A(1, 3)$ is 1 unit right and 3 units up from the origin.

$B(–1, 3)$ is 1 unit left and 3 units up from the origin.

$C(–2, –4)$ is 2 units left and 4 units down from the origin.

$D(3, –4)$ is 3 units right and 4 units down from the origin.

EXAMPLE 5

State which point from Example 4 lies in:

a the 3rd quadrant **b** the 1st quadrant **c** the 4th quadrant.

SOLUTION

a C **b** A **c** D

1 In which quadrant does the point (–1, –5) lie?

2 Where is the point (0, 3) positioned on the number plane? Select the correct answer **A, B, C** or **D**.

   A *x*-axis   **B** *y*-axis   **C** 1st quadrant   **D** the origin

3 Copy and complete this paragraph.

   On the number plane, the _____ axis is called the *x*-axis and the vertical axis is called the __ axis. To plot a point, start at the origin and move left or _____ first for the *x*-value and then up or _____ for the *y*-value.

4 **a** Plot each point below on a number plane.

   A (3, 2)   **B** (–1, 4)   **C** (–3, –1)   **D** (4, 2)   **E** (–1, 5)

   b State which quadrant each point lies in.

5 For the number plane below, write the coordinates of points *A* to *J*.

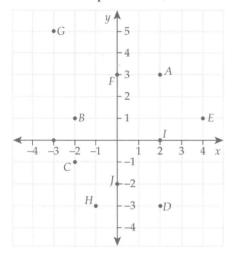

6 State which points from Question **5** lie:

   a in the 2nd quadrant   **b** in the 3rd quadrant

   c on the *y*-axis   **d** in the 1st quadrant

7 Plot and join each set of points on a separate number plane, then state what shape is formed.

   a (0, 2)   (–3, –2)   (3, –2)

   b (0, 1)   (–2, –2)   (3, –2)   (5, 1)

   c (0, –3)   (3, 0)   (0, 3)   (–3, 0)

   d (2, 3)   (–2, 3)   (–4, 0)   (–2, –3)   (2, –3)   (4, 0)

Each column in a table of values represents the coordinates of a point on the number plane. For example, this column represents the point (2, 5).

x	2
y	5

EXAMPLE 6

Graph this table of values on a number plane.

x	-2	-1	0	1	2
y	-4	-2	0	2	4

SOLUTION

Reading the table of values in columns, we get the coordinates of the points.

(−2, −4)   (−1, −2)   (0, 0)   (1, 2)   (2, 4)

Plotting these points on the number plane:

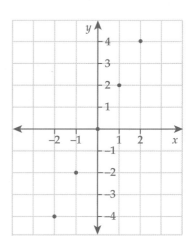

EXAMPLE 7

Graph this table of values after completing it.

$y = 6 - 3x$

x	-1	0	1	2
y				

SOLUTION

x	-1	0	1	2
y	9	6	3	0

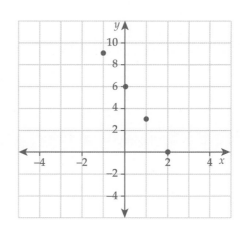

ISBN 9780170351058

1 If $y = 2x + 1$, find y when $x = -3$. Select the correct answer **A**, **B**, **C** or **D**.

  A 3   **B** –7   **C** –5   **D** –6

2 Which point below lies on $y = 4 - 2x$? Select **A**, **B**, **C** or **D**.

  A (–1, 6)   **B** (1, –2)   **C** (–2, 0)   **D** (–3, –2)

3 Graph each table of values on a separate number plane.

a

x	−2	−1	0	1	2
y	−1	0	1	2	3

b

x	−2	−1	0	1	2
y	−6	−3	0	3	6

c

x	−2	−1	0	1	2
y	−4	−3	−2	−1	0

d

x	−2	−1	0	1	2
y	0	1	2	3	4

e

x	−2	−1	0	1	2
y	4	2	0	−2	−4

f

x	−3	−1	0	2	3
y	−7	−3	−1	3	5

4 a What do you notice about each set of points you graphed in Question 3?

  b Could the pattern be continued?

5 Copy and complete each table of values and then graph the values on a number plane.

a $y = x + 5$

x	−4	−2	0	1	2
y	1		5		

b $y = 3x - 2$

x	−2	−1	0	1	2
y		−5			

c $y = 5 - x$

x	−2	−1	0	1	2
y	7				

d $y = 4 - 2x$

x	−3	−1	0	2	3
y					

e $y = 4x - 10$

x	−1	0	1	2
y				

f $y = 5x - 3$

x	1	0	3	2
y				

WORDBANK

interval A section of a line with a definite length.

An interval can be drawn to join two points on the number plane, and we can use Pythagoras' theorem to find the length of the interval.

PYTHAGORAS' THEOREM

If the lengths of the sides in a right-angled triangle are a, b and c, where c is the length of the hypotenuse (longest side), then
$c^2 = a^2 + b^2$

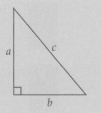

EXAMPLE 8

Find the length of the interval joining each pair of points.

a (1, 4) and (5, 1) **b** (−2, 2) and (6, 5)

SOLUTION

Plot the points on the number plane first, then draw an interval between them.

Then draw a right-angled triangle using the interval as the hypotenuse.

a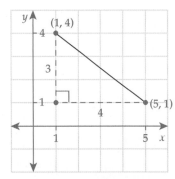

b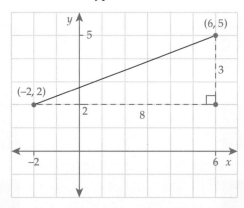

Next, find the lengths of the vertical and horizontal sides of the triangle by counting squares.

Then use Pythagoras' theorem to find the length of the interval.

$c^2 = a^2 + b^2$
$\quad = 3^2 + 4^2$
$\quad = 9 + 16$
$\quad = 25$
$c = \sqrt{25}$
$c = 5$

$c^2 = a^2 + b^2$
$\quad = 8^2 + 3^2$
$\quad = 64 + 9$
$\quad = 73$
$c = \sqrt{73}$
or $c \approx 8.54$

ISBN 9780170351058

Extension: The distance formula

The formula for Pythagoras' theorem can be modified to find the length of an interval (the distance, d) joining points (x_1, y_1) and (x_2, y_2). This 'distance formula' is more complicated but can be used instead of drawing a diagram.

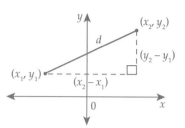

$$d = \sqrt{(x_2 - x_1)^2 + (y_2 - y_1)^2}$$

Length of the horizontal line   Length of the vertical line

EXAMPLE 9

Use the distance formula to find, correct to one decimal place, the length of the interval joining each pair of points.

a $(3, 5)$ and $(6, -4)$ **b** $(-2, 8)$ and $(5, -6)$

SOLUTION

a $x_1 = 3,\ y_1 = 5,\ x_2 = 6,\ y_2 = -4$

$d = \sqrt{(x_2 - x_1)^2 + (y_2 - y_1)^2}$

$\quad = \sqrt{(6 - 3)^2 + (-4 - 5)^2}$

$\quad = \sqrt{3^2 + (-9)^2}$

$\quad = \sqrt{9 + 81}$

$\quad = \sqrt{90}$

$\quad \approx 9.5$

b $x_1 = -2,\ y_1 = 8,\ x_2 = 5,\ y_2 = -6$

$d = \sqrt{(x_2 - x_1)^2 + (y_2 - y_1)^2}$

$\quad = \sqrt{[5 - (-2)]^2 + (-6 - 8)^2}$

$\quad = \sqrt{7^2 + (-14)^2}$

$\quad = \sqrt{49 + 196}$

$\quad = \sqrt{245}$

$\quad \approx 15.7$

EXERCISE 15-05

1 Which is the correct formula for Pythagoras' theorem, if c is the hypotenuse? Select the correct answer **A**, **B**, **C** or **D**.

   A $c^2 = b^2 - a^2$ **B** $a^2 = b^2 + c^2$

   C $c^2 = a^2 - b^2$ **D** $c^2 = a^2 + b^2$

2 Find the length of the hypotenuse if the other two sides of the right-angled triangle are 5 and 12. Select **A**, **B**, **C** or **D**.

   A   169   **B**   13   **C** $\sqrt{119}$ **D** $\sqrt{-119}$

3 Is each statement true or false?

  a You will always get a whole number when you find the length of an interval.

  b You can only use one method to find the length of an interval.

  c You can write an exact answer or round it to a particular number of decimal places.

  d The length of an interval may be positive or negative.

4 Copy each interval, draw right-angled triangles and use Pythagoras' theorem to find the length of the interval in exact form (not rounded).

a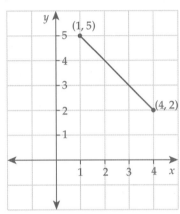

b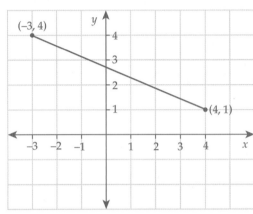

c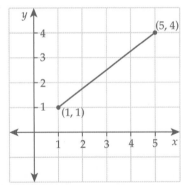

d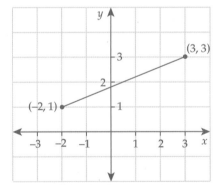

5 Find (correct to one decimal place if necessary) the length of the interval joining each pair of points.

  a (3, 1) and (6, 9)   b (2, 4) and (5, 8)

  c (−1, 6) and (8, −4)   d (−5, −2) and (7, 4)

  e (3, 6) and (−4, −2)   f (−4, 1) and (2, 9)

  g (−4, −6) and (1, 7)   h (−5, 6) and (7, 2)

  i (−1, 3) and (6, −2)   j (−5, −1) and (8, 4)

15-06 | The midpoint of an interval

WORDBANK

midpoint of an interval   The point at the middle of an interval.

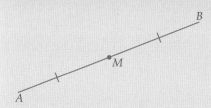

To find the midpoint of the interval joining points (x_1, y_1) **and** (x_2, y_2)**:**

- find the average of the x-values: $\dfrac{x_1 + x_2}{2}$

- find the average of the y-values: $\dfrac{y_1 + y_2}{2}$

- the midpoint is $\left(\dfrac{x_1 + x_2}{2}, \dfrac{y_1 + y_2}{2} \right)$.

EXAMPLE 10

Find the midpoint of the interval joining each pair of points.

a $(5, -2)$ and $(1, 6)$ **b** $(-2, 2)$ and $(4, 5)$

SOLUTION

a $x_1 = 5, y_1 = -2, x_2 = 1, y_2 = 6$

$\dfrac{x_1 + x_2}{2} = \dfrac{5 + 1}{2} = \dfrac{6}{2} = 3$

$\dfrac{y_1 + y_2}{2} = \dfrac{-2 + 6}{2} = \dfrac{4}{2} = 2$

Midpoint is $(3, 2)$.

b $x_1 = -2, y_1 = 2, x_2 = 4, y_2 = 5$

$\dfrac{x_1 + x_2}{2} = \dfrac{-2 + 4}{2} = \dfrac{2}{2} = 1$

$\dfrac{y_1 + y_2}{2} = \dfrac{2 + 5}{2} = \dfrac{7}{2} = 3\dfrac{1}{2}$

Midpoint is $(1, 3\dfrac{1}{2})$.

These answers can also be read from the graphs of the intervals.

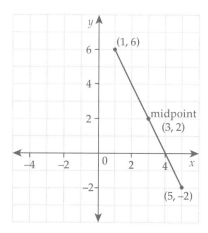

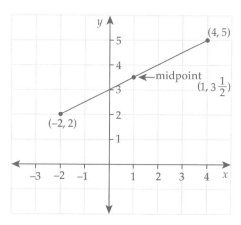

1   What is the average of 4 and 8? Select the correct answer **A**, **B**, **C** or **D**.

    A 5   **B** 6   **C** 7   **D** 8

2   What is the midpoint of the interval joining (4, 2) and (8, 6)? Select **A**, **B**, **C** or **D**.

    A (6, 2)   **B** (5, 4)   **C** (6, 4)   **D** (7, 3)

3   Is each statement true or false?

    a The average of 5 and 7 is 6.

    b The average of 6 and 10 is 7.

    c The average of –4 and –6 is –5.

    d The average of –2 and 6 is 2.

4   **a** What is the average of 5 and 9?

    b What is the average of 1 and 7?

    c What is the midpoint of (5, 1) and (9, 7)?

5   Write the coordinates of the midpoint of each interval.

    a   **b**

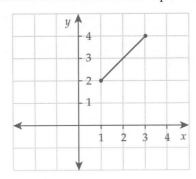   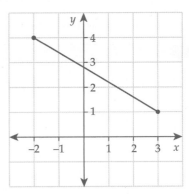

6   By drawing a diagram, find the midpoint of the interval joining (–4, 0) and (0, 8).

7   Find the midpoint of the interval joining each pair of points.

    a (1, 4) and (3, 6)   **b** (2, 6) and (6, 10)

    c (3, 8) and (5, 12)   **d** (7, 9) and (3, 12)

    e (–1, 5) and (5, 9)   **f** (2, –4) and (6, –8)

    g (–3, 5) and (4, –3)   **h** (–1, 8) and (–3, –6)

    i (–2, –8) and (–4, 6)   **j** (4, –10) and (–6, 8)

8   Find the other endpoint of an interval with:

    a one endpoint (2, –4) and a midpoint of (4, 6).

    b one endpoint (–3, 5) and a midpoint of (2, 1).

WORDBANK

gradient   A value that measures the slope or steepness of an interval or line.

$$\text{Gradient} = \frac{\text{rise}(\uparrow)}{\text{run}(\rightarrow)}$$

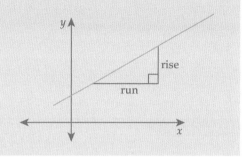

EXAMPLE 11

Find the gradient of the interval joining each pair of points.

a   (2, 1) and (5, 8)   b   (−1, 5) and (2, 3)

SOLUTION

Plot the points on the number plane first, then draw an interval between them.

Next draw a right-angled triangle using the interval as the hypotenuse.

a

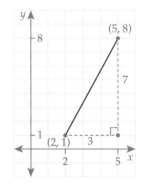

$$\text{Gradient} = \frac{\text{rise}}{\text{run}}$$
$$= \frac{7}{3}$$

b

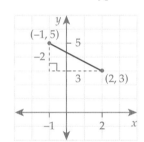

✳ A 'drop' of 2 means a rise of −2.

$$\text{Gradient} = \frac{\text{rise}}{\text{run}}$$
$$= \frac{-2}{3}$$
$$= -\frac{2}{3}$$

$$\text{Gradient} = \frac{\text{rise}}{\text{run}}$$

■ A line **increasing** from left to right has a **positive** gradient.
■ A line **decreasing** from left to right has a **negative** gradient.

Extension: The gradient formula

The formula for the gradient, m, of the line or interval joining points (x_1, y_1) and (x_2, y_2) is more complicated but can be used instead of drawing a diagram.

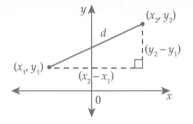

$$m = \frac{\text{rise}}{\text{run}} = \frac{y_2 - y_1}{x_2 - x_1}$$

The rise is the difference in y-values.
The run is the difference in x-values.

EXAMPLE 12

Use the gradient formula to find the gradient of the line joining each pair of points (same points as in Example 10).

a (2, 1) and (5, 8)

b (–1, 5) and (2, 3)

SOLUTION

a $x_1 = 2, y_1 = 1, x_2 = 5, y_2 = 8$

$$m = \frac{y_2 - y_1}{x_2 - x_1}$$

$$= \frac{8 - 1}{5 - 2}$$

$$= \frac{7}{3}$$

b $x_1 = -1, y_1 = 5, x_2 = 2, y_2 = 3$

$$m = \frac{y_2 - y_1}{x_2 - x_1}$$

$$= \frac{3 - 5}{2 - (-1)}$$

$$= \frac{-2}{3}$$

$$= -\frac{2}{3}$$

EXERCISE 15-07

1 What is the vertical rise from (–1, 5) to (2, 9)? Select the correct answer **A, B, C** or **D**.

    A 2   **B** 3   **C** 4   **D** 5

2 What is the gradient of the line joining (–1, 5) to (2, 9)? Select **A, B, C** or **D**.

    A 4   **B** $\frac{4}{3}$ **C** $\frac{3}{4}$ **D** 2

3 Is each statement about the gradient true or false?

    a It measures steepness.   **b** It is always positive.

    c It is the run divided by the rise.   **d** It may be positive, negative or zero.

4 **a** What is the rise from (–2, 4) to (5, 11)?

    b What is the run from (–2, 4) to (5, 11)?

    c What is the gradient of the line joining (–2, 4) to (5, 11)?

5 Find the gradient of each interval.

a

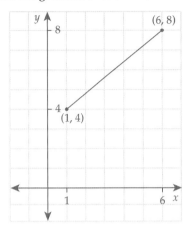

b

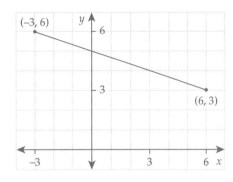

c

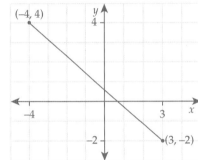

d

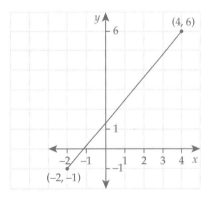

6 Find the gradient of the line joining each pair of points.

    a (3, 5) and (8, 10)   **b** (−2, 4) and (6, −1)

    c (−3, 7) and (4, 6)   **d** (−1, 2) and (5, 9)

    e (2, 6) and (4, −2)   **f** (−2, 1) and (6, 11)

    g (−5, −4) and (−2, 1)   **h** (−2, 4) and (3, 9)

    i (6, −1) and (−2, 5)   **j** (−4, 8) and (3, 12)

CROSSWORD PUZZLE

Print a copy of this puzzle, then complete this crossword using the clues below.

Across

2   The point halfway between two other points.

5   One quarter of a number plane.

7   Is the x-axis vertical or horizontal?

10 The slope of a line.

Down

1   Points on a number plane can be located using their x- and y-_____.

3   Is the y-axis vertical or horizontal?

4   The distance formula can be used to calculate the _____ of an interval.

6   The horizontal difference between two points on an interval, used for calculating gradient.

8   The vertical difference between two points on an interval, used for calculating gradient.

9   We can use a rule or formula to complete and graph a _____ of values.

PRACTICE TEST 15

Part A General topics

Calculators are not allowed.

1 Convert 12% to a simple fraction.

2 Expand $3a(2a - 8)$.

3 Evaluate $24 + 72 ÷ 9 × (-4)$

4 How many minutes are there from 3:15 p.m. to 4:55 p.m.?

5 Find an algebraic expression for the perimeter of this figure.

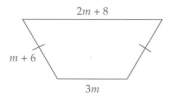

6 Write 1 025 920 correct to three significant figures.

7 Find the median of these scores.
   12, 5, 12, 18, 11, 20, 17, 3.

8 If the probability that you will live past age 100 is 9%, what is the probability that you won't live past 100?

9 Copy and complete: $1 \text{ m}^3 = $ _____ L.

10 Find the value of y.

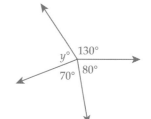

Part B Coordinate geometry

Calculators are allowed.

15–01 Tables of values

11 Find the value of y when $x = -1$ if $y = 3x + 8$. Select the correct answer **A**, **B**, **C** or **D**.

   A 9   **B** 5   **C** 11   **D** 7

12 Complete each table of values.

   a $y = x - 3$

x	6	4	2
y			

   b $y = 2x + 3$

x	-1	0	1
y			

15–02 Finding the rule

13 Find the rule connecting x and y for this table of values.

x	-2	-1	0	1	2
y	-3	-1	1	3	5

15–03 The number plane

14 In which quadrant does the point $(3, -4)$ lie?

15 Plot each point on a number plane.

   A $(1, 4)$ **B** $(-2, 3)$ **C** $(5, -2)$ **D** $(-3, -4)$ **E** $(-2, 0)$

15–04 Graphing tables of values

16 Graph this table of values on a number plane.

x	−2	−1	0	1	2
y	−5	−3	−1	1	3

17 Copy and complete this table of values, then graph the values on a number plane.

$y = 6 - 2x$

x	−1	0	1	2
y				

15–05 The length of an interval

18 Find the exact length of this interval.

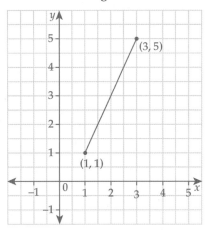

15–06 The midpoint of an interval

19 Find the midpoint of the interval in Question **18**.

20 Find the midpoint joining the points (−2, 3) and (4, −9).

15–07 The gradient of a line

21 Find the gradient of the interval in Question **18**.

22 Find the gradient of the line joining the points (4, −3) and (7, −6).

GRAPHING LINES AND CURVES

16

IN THIS CHAPTER YOU WILL:

- graph linear equations on the number plane
- test if a point lies on a line
- find the equation of horizontal and vertical lines
- find the equation of a line
- graph parabolas with equations of the form $y = ax^2 + c$
- graph exponential equations
- graph circles

Shutterstock.com/sainthorant daniel

WORDBANK

linear equation An equation that connects two variables, usually x and y, whose graph is a straight line.

x-intercept The value where a line crosses the x-axis.

y-intercept The value where a line crosses the y-axis.

To graph a linear equation:
- ▇ complete a table of values using the equation
- ▇ plot the points from the table on a number plane
- ▇ join the points to form a straight line.

To be sure, it is best to find three points on the line using a table of values.

We can substitute any x-values into the linear equation, but $x = 0$, $x = 1$ or $x = 2$ are usually the easiest to use.

EXAMPLE 1

Graph each linear equation on a number plane.

a $y = 2x + 1$ **b** $y = 4 - 3x$

SOLUTION

Complete a table of values for each linear equation.

a $y = 2x + 1$

x	0	1	2
y	1	3	5

b $y = 4 - 3x$

x	0	1	2
y	4	1	-2

Graph the table of values, rule the line and label it with the equation.

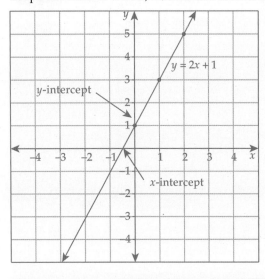

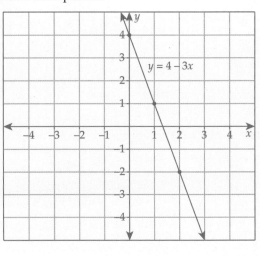

Draw arrows on the ends of the line because a line has an infinite number of points and goes on endlessly in both directions.

Graphing linear equations

EXAMPLE 2

What are the x and y-intercepts of the line with equation $y = 2x + 1$ in Example **1**?

SOLUTION

Read off the intercepts from the graphs.

The line crosses the x-axis at $-\frac{1}{2}$, so its x-intercept is $-\frac{1}{2}$.

The line crosses the y-axis at 1, so its y-intercept is 1.

EXERCISE 16-01

1 How many points are best for graphing a linear equation? Select the correct answer **A, B, C** or **D**.

   A 1   **B** 2   **C** 3   **D** 4

2 Which of these points lie on the line $y = 3x - 1$? Select **A, B, C** or **D**.

   A (−1, −2)   **B** (2, −5)   **C** (−1, 1)   **D** (0, −1)

3 Write down the x- and y-intercepts of each line below.

   a
   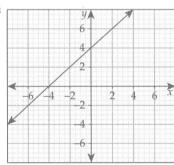

   b
   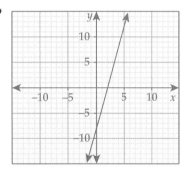

   c
   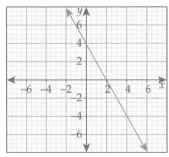

   d
   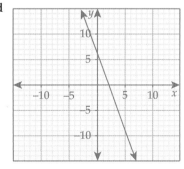

4 Graph each linear equation on a number plane after completing a table of values, and state the x-intercept and y-intercept of each line.

   a $y = 3x$ **b** $y = x + 2$ **c** $y = 4 - x$ **d** $y = 4x - 1$

   e $y = 8 - 2x$ **f** $y = -2x + 4$ **g** $y = -2x + 6$ **h** $y = -2x - 1$

> **To test if a point lies on a line:**
> - substitute the x-value and the y-value into the linear equation
> - if LHS = RHS, then the point lies on the line
> - otherwise, the point does not lie on the line.
>
> A point must satisfy the equation of the line to lie on the line.

EXAMPLE 3

Test whether each point lies on the line with equation $y = 2x + 3$.

a $(1, 5)$ **b** $(2, 4)$ **c** $(-1, 1)$

SOLUTION

Substitute each point into the equation $y = 2x + 3$.

a For $(1, 5)$, $x = 1$, $y = 5$

$y = 2x + 3$

$5 = 2 \times 1 + 3$

$5 = 5$ True

So $(1, 5)$ lies on the line.

b For $(2, 4)$, $x = 2$, $y = 4$

$y = 2x + 3$

$4 = 2 \times 2 + 3$

$4 = 7$ False

So $(2, 4)$ does *not* lie on the line.

c For $(-1, 1)$, $x = -1$, $y = 1$

$y = 2x + 3$

$1 = 2 \times (-1) + 3$

$1 = 1$ True

So $(1, 1)$ lies on the line.

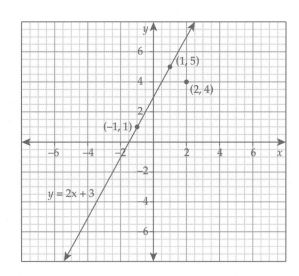

1   Which point lies on the line with equation $y = 3x - 2$? Select the correct answer **A**, **B**, **C** or **D**.

   A $(1, -1)$ **B** $(-1, 5)$ **C** $(-2, -8)$ **D** $(2, 5)$

2   Which of these lines has the point $(-2, 1)$? Select **A**, **B**, **C** or **D**.

   A $y = 2x - 1$ **B** $y = 2x + 5$ **C** $y = 2x + 1$ **D** $y = 2x - 3$

3   Copy and complete these sentences.

   To test if a point lies on a line, substitute the x-value and _____ into the linear equation.

   If the equation is true, then the point ____ on the line. If it is _____ the point does not ___ on the line.

4   Which of these points lie on the line $y = 4x + 2$?

   a $(1, 3)$ **b** $(-1, -2)$ **c** $(2, 6)$ **d** $(-2, -10)$ **e** $(0, 2)$

5   Graph $y = 4x + 2$ on a number plane and check your answers to Question **4**.

6   Which of these points lie on the line $y = 8 - 2x$?

   a $(1, 6)$ **b** $(-1, 6)$ **c** $(2, 4)$ **d** $(-2, -4)$ **e** $(0, -8)$

7   For each linear equation, test whether the given points lie on the line.

   a $y = 2x - 4$ $(2, -3), (-1, -6)$

   b $y = 12 - x$ $(-2, 14), (4, 3)$

   c $x + y = 7$ $(4, -3), (3, -4)$

   d $2x + y = 5$ $(-1, 7), (1, -3)$

   e $y = 4x - 6$ $(4, 10), (2, 0)$

   f $2x - 3y = 6$ $(2, 1), (-3, -4)$

   g $y = 3x - 1$ $(0, -1), (-1, 4)$

   h $x + 2y = 8$ $(1, 4), (-2, 5)$

8   Without graphing the line, give the coordinates of a point that lies on the line with equation $y = 3x - 4$.

Anstock / Alamy

WORDBANK

horizontal A line that is flat, parallel to the horizon.

vertical A line that is straight up and down, at right angles to the horizon.

constant A number, not a variable.

A **horizontal line** has equation $y = c$, where c is a constant (number).
A **vertical line** has equation $x = c$, where c is a constant (number).

EXAMPLE 4

Graph each line on a number plane.

a $y = 3$ **b** $y = -2$ **c** $x = 2$ **d** $x = -3$

SOLUTION

a $y = 3$ is a horizontal line with a
y-intercept of 3.

All points on $y = 3$ have a y-value of 3,
such as $(-1, 3)$ and $(1, 3)$.

b $y = -2$ is a horizontal line with a
y-intercept of -2.

All points on $y = -2$ have a y-value
of -2, such as $(-2, -2)$ and $(2, -2)$.

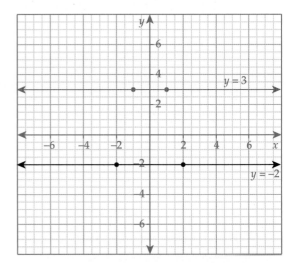

c $x = 2$ is a vertical line with an
x-intercept of 2.

All points on $x = 2$ have a x-value of 2,
such as $(2, -2)$ and $(2, 2)$.

d $x = -3$ is a vertical line with an
x-intercept of -3.

All points on $x = -3$ have a
x-value of -3, such as $(-3, -2)$ and $(-3, 2)$.

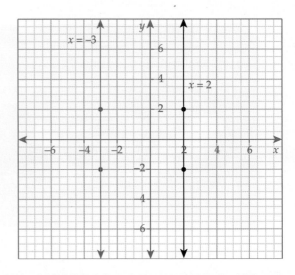

1   What types of lines are the graphs of $x = 3$ and $y = -1$? Select the correct answer **A**, **B**, **C** or **D**.

    A Both horizontal   **B** $x = 3$ vertical and $y = -1$ horizontal

    C Both vertical   **D** $x = 3$ horizontal and $y = -1$ vertical

2   What is the point of intersection of the lines $x = 3$ and $y = -1$? Select **A**, **B**, **C** or **D**.

    A $(1, -3)$ **B** $(-3, -1)$ **C** $(-1, 3)$ **D** $(3, -1)$

3   Is each statement true or false?

    a $x = 5$ is a horizontal line.

    b $y = -5$ is a horizontal line.

    c $x = -3$ is a vertical line.

    d $y = 3$ is a vertical line.

4   Write down the equation of each line shown.

    a 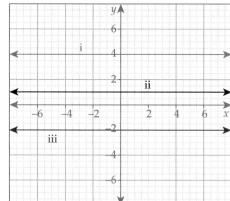   **b**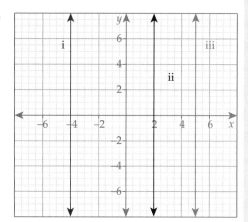

5   Graph each line on the same number plane.

    a $x = -1$ **b** $x = 0$ **c** $x = 3$ **d** $x = 4$

6   Graph each line on the same number plane.

    a $y = -4$ **b** $y = -3$ **c** $y = 0$ **d** $y = 1$

7   Write down the equation of a line that is

    a horizontal with a y-intercept of 4

    b vertical with an x-intercept of -6

    c horizontal, passing through $(-2, 4)$

    d vertical, passing through $(5, -7)$.

8   Find the point of intersection of each pair of lines.

    a $x = 8$ and $y = -4$ **b** $x = -7$ and $y = 6$.

There are two methods for finding the equation of a line, using skills you learnt in Chapter 15:
- using a table of values
- using the gradient and y-intercept

EXAMPLE 5

Find the equation of this line.

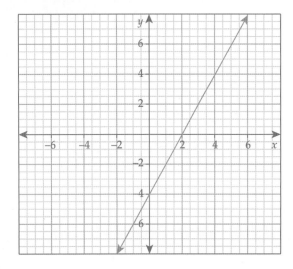

SOLUTION

Method 1: Table of values

Use points on the line to create a table of values with x-values that increase by 1.

x	0	1	2
y	−4	−2	0

$+2$ $+2$

y increases by 2 each time, so the multiplier is 2.

So the rule is: $y = 2x$ _____.

Using the columns in the table:

$2 \times 0 - 4 = -4$

$2 \times 1 - 4 = -2$, and so on.

So the equation is: $y = 2x - 4$.

ISBN 9780170351058

Finding the equation of a line

Method 2: Gradient and y-intercept

Choose two points on the line to find the gradient, say $(0, -4)$ and $(2, 0)$.

Gradient $m = \dfrac{4}{2} = 2$, so the multiplier is 2.

From the graph, the y-intercept is -4, so the equation of the line is $y = 2x - 4$.

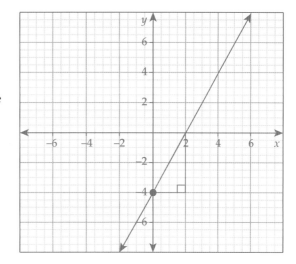

EXAMPLE 6

Find the equation of this line.

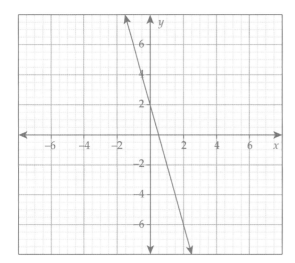

SOLUTION

x	0	1	2
y	2	-2	-6

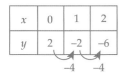

y decreases by 4 each time, so the multiplier is -4.

So the rule is $y = -4x$ _____.

$-4 \times 0 + 2 = 2$

$-4 \times 1 + 2 = -2$, and so on.

So the equation is $y = -4x + 2$.

1   What is the equation of a line where each y-value is 3 less than the x-value? Select the correct answer **A, B, C** or **D**.

   A $y = x + 3$ **B** $y = 3 - x$ **C** $y = x - 3$ **D** $x = y - 3$

2   What is the equation of the line with a gradient of 3 and a y-intercept of –2? Select **A, B, C** or **D**.

   A $y = -2x + 3$ **B** $y = 3x - 2$ **C** $y = 3 - 2x$ **D** $y = -3x - 2$

3   Find the equation of each table of values.

a

x	-1	0	1
y	8	9	10

b

x	-1	0	1
y	-3	0	3

c

x	-1	0	1
y	-2	1	4

d

x	-1	0	1
y	-3	-1	1

4   **a**   Find the gradient of the line below.

   b   What is the y-intercept of the line?

   c   Write down the equation of this line.

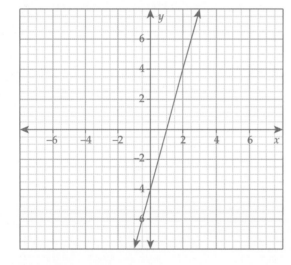

5   Find the equation of each line.

a

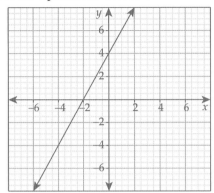

b

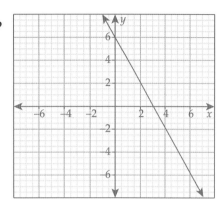

c

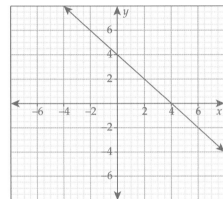

d

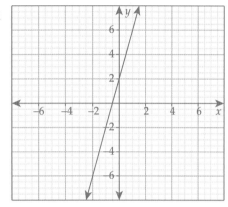

e

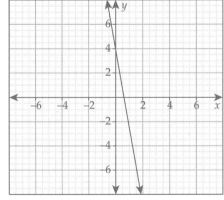

f

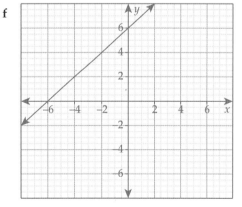

6   a   Graph the line $y = 10 - 2x$.

    b   What is the gradient and y-intercept of this line?

WORDBANK

quadratic equation An equation involving x^2, such as $y = x^2 - 5$, $y = 3x^2 + 2$ and $y = \frac{1}{2}x^2$.

parabola The graph of a quadratic equation, a U-shaped curve.

To graph a parabola:
- complete a table of values using the equation
- plot the points from the table on a number plane
- join the points to form a smooth curve.

To be sure, it is best to find many points on the curve using a table of values.

We can substitute any x-values into the linear equation but values between −2 and 2 are usually easiest to use.

EXAMPLE 7

Graph the quadratic equation $y = x^2$ on the number plane.

SOLUTION

$y = x^2$

x	−3	−2	−1	0	1	2	3
y	9	4	1	0	1	4	9

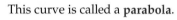
$(-3)^2 \quad (-2)^2 \quad (-1)^2 \quad 0^2 \quad 1^2 \quad 2^2 \quad 3^2$

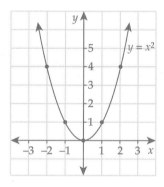

This curve is called a **parabola**.

All parabolas have this same basic shape.

EXAMPLE 8

Graph each quadratic equation on a number plane.

a $y = 2x^2$ **b** $y = -x^2$

SOLUTION

a $y = 2x^2$

x	−2	−1	0	1	2
y	8	2	0	2	8

This parabola is steeper than the parabola for $y = x^2$.
This parabola is **concave up**.

When a **positive** number is in front of x^2, the parabola is concave up.

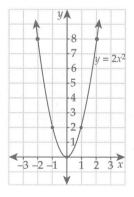

b $y = -x^2$

x	-2	-1	0	1	2
y	-4	-1	0	-1	-4

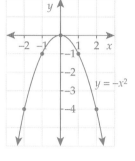

This parabola is the parabola $y = x^2$ turned upside-down.
This parabola is **concave down**.

When a **negative** number is in front of x^2, the parabola is concave down.

EXAMPLE 9

Graph each parabola.

a $y = x^2 + 1$

b $y = -x^2 - 1$

SOLUTION

a

x	-2	-1	0	1	2
y	5	2	1	2	5

b

x	-2	-1	0	1	2
y	-5	-2	-1	-2	-5

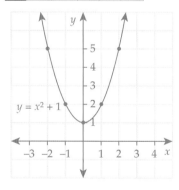

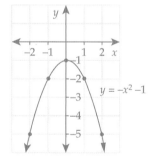

This parabola is $y = x^2$ moved up 1 unit so that its y-intercept is 1.

This parabola is the parabola $y = -x^2$ moved down 1 unit so that its y-intercept is -1.

This parabola is **concave up**.

This parabola is **concave down**.

1   Which of these is a quadratic equation, the equation of a parabola? Select the correct answer **A**, **B**, **C** or **D**.

   A $y = 2x - 1$ **B** $y = 1 - 2x$ **C** $y = 1 + 2x$ **D** $y = x^2 - 1$

2   What is the equation of a parabola that is concave down with a y-intercept of 2? Select the correct answer **A**, **B**, **C** or **D**.

   A $y = x^2 + 2$ **B** $y = -x^2 + 2$ **C** $y = -x^2 - 2$ **D** $y = 2x^2$

3   Graph each parabola on the same number plane.

   a $y = x^2$ **b** $y = 2x^2$ **c** $y = -x^2$ **d** $y = -2x^2$

4   Which parabolas in Question **3** are concave up?

5   Graph each set of parabolas on a separate number plane. Check that each graph is correct by using graphing software such as GeoGebra.

   a $y = x^2, y = x^2 + 2, y = x^2 + 4$
   b $y = x^2, y = x^2 - 1, y = x^2 - 4$
   c $y = -x^2, y = -x^2 + 2, y = 4 - x^2$
   d $y = x^2, y = \frac{1}{2}x^2, y = \frac{1}{4}x^2$
   e $y = 2x^2 + 4, y = \frac{1}{2}x^2 + 4$
   f $y = -2x^2 - 1, y = -\frac{1}{2}x^2 - 1$

Alamy/Martin Grace

WORDBANK

exponent Another name for power. For example, 2^3 has a power or exponent of 3.

exponential equation An equation with x as the exponent or power, such as $y = 2^x$.

asymptote A line that a curve approaches but never meets.

EXAMPLE 10

Graph each exponential equation on the number plane.

a $y = 2^x$ **b** $y = 2^{-x}$

SOLUTION

a $y = 2^x$

x	−2	−1	0	1	2
y	$\frac{1}{4}$	$\frac{1}{2}$	1	2	4

$\qquad 2^{-2} \quad 2^{-1} \quad 2^0 \quad 2^1 \quad 2^2$

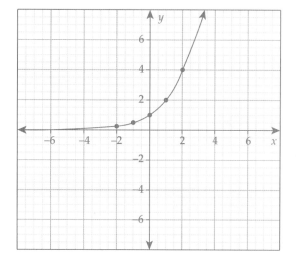

Note that the y-values are doubling.

The graph of an exponential equation increases slowly when x is small but increases more quickly as x becomes larger in value.

For small values of x, the curve appears to touch the x-axis, but it doesn't: it just gets closer and closer. The x-axis is called an **asymptote** of the graph.

b $y = 2^{-x}$

x	−2	−1	0	1	2
y	4	2	1	$\frac{1}{2}$	$\frac{1}{4}$

$\qquad 2^2 \quad 2^1 \quad 2^0 \quad 2^{-1} \quad 2^2$

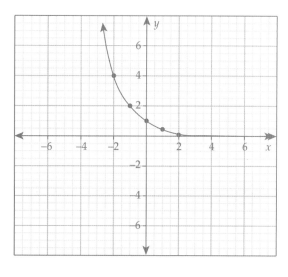

Note that the y-values are being halved.

The graph of an exponential equation with a *negative* power decreases quickly as x is small but decreases more slowly as x becomes larger in value.

For large values of x the curve appears to touch the x-axis, but the x-axis is still an **asymptote** of the graph.

1   Which of these is an exponential equation? Select the correct answer **A, B, C** or **D**.

   A $y = 2x$ **B** $y = \dfrac{2}{x}$ **C** $y = -2x^2$ **D** $y = 2^x$

2   Graph each exponential equation on the same number plane.

   a $y = 3^x$ **b** $y = 4^x$ **c** $y = 3^{-x}$ **d** $y = 4^{-x}$

3   Which curves in Question **2** are increasing?

4   Which phrase is true about the graph of $y = 2^{-x}$? Select **A, B, C** or **D**.

   A increases as x increases   **B** decreases as x increases

   C the y-axis as an asymptote   **D** does not have an asymptote

5   Is each statement true or false?

   a $y = 3^x$ is an exponential equation.

   b $y = \dfrac{3}{x}$ is not an exponential equation.

   c An exponential graph will always have an asymptote.

   d The x-axis is an asymptote of $y = 3^x$.

   e $y = 3^x$ is a decreasing curve.

   f $y = 3^{-x}$ is a decreasing curve.

6   Graph each set of exponential equations on a separate number plane. Check that each graph is correct using graphing software such as GeoGebra.

   a $y = 2^x, y = 4^x, y = 6^x$ **b** $y = 2^{-x}, y = 4^{-x}, y = 6^{-x}$

   c $y = 3^x, y = 3^{-x}, y = \left(\dfrac{1}{3}\right)^x$ **d** $y = 2^x + 1, y = 2^x - 2$

   e $y = 2^{-x} + 3, y = 2^{-x} - 1$

7   Write down the asymptotes for the exponential graphs in questions **6c** and **6e**.

Shutterstock.com/Pi-Lens

16-07 Graphing circles

The **equation of a circle** with centre $(0, 0)$ and radius r units is $x^2 + y^2 = r^2$.

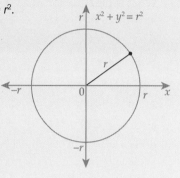

EXAMPLE 11

Graph the circle $x^2 + y^2 = 4$.

SOLUTION

The radius is r, where $r^2 = 4$.

$r = \sqrt{4} = 2$.

A circle with centre $(0, 0)$ and radius 2 units.

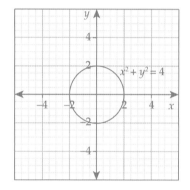

EXAMPLE 12

Find the equation of this circle.

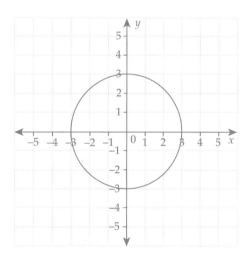

SOLUTION

The radius is of this circle is 3, so $r = 3$, and the equation of the circle is:

$x^2 + y^2 = 3^2$

$x^2 + y^2 = 9$

1   What is the equation of a circle with centre (0, 0) and radius 6 units? Select the correct answer **A**, **B**, **C** or **D**.

    A $x^2 + y^2 = 6$ **B** $x^2 + y^2 = 36$

    C $x^2 + y^2 = \sqrt{6}$ **D** $x^2 - y^2 = 36$

2   Graph each circle on a number plane.

    a $x^2 + y^2 = 9$ **b** $x^2 + y^2 = 25$

    c $x^2 + y^2 = 49$ **d** $x^2 + y^2 = 4$

    e $x^2 + y^2 = 16$ **f** $x^2 + y^2 = 64$

    g $x^2 + y^2 = 5$ **h** $x^2 + y^2 = 13$

    i $x^2 + y^2 = 35$

3   What is the equation of a circle with centre (0, 0) and radius 4 units? Select **A**, **B**, **C** or **D**.

    A $x^2 + y^2 = 16$ **B** $x^2 + y^2 = 4$

    C $x^2 + y^2 = 2$ **D** $x^2 + y^2 = -4$

4   Write down the equation of each circle with centre (0, 0) and:

    a radius 1 unit   **b** radius 9 units

    c radius 6 units   **d** diameter 6 units

    e diameter 14 units   **f** radius 5 units

    g radius 11 units   **h** diameter 16 units

    i diameter 20 units

5   Write down the equation of each circle.

    a

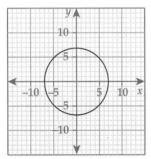

    b

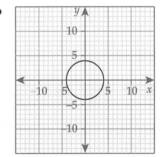

CODE PUZZLE

Match each graph to its equation on the next page, and use matched numbers and letters to solve the riddle on the next page.

1
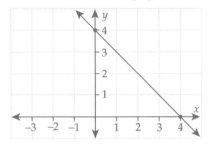

2
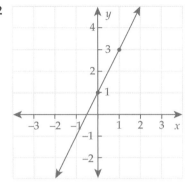

3
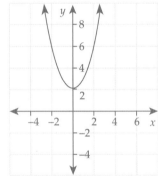

4
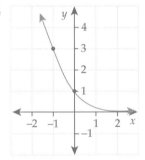

5
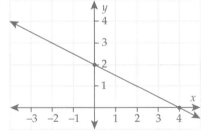

6
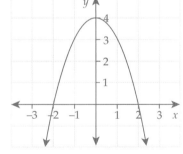

7
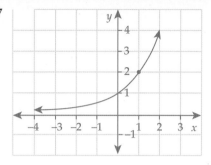

8
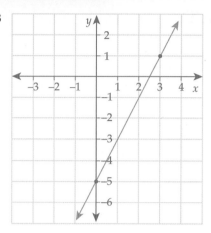

9
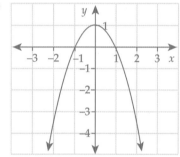

A. $y = 1 - x^2$ **B.** $y = 2^x$ **D.** $x + 2y = 4$

E. $y = 2x + 1$ **I.** $y = x^2 + 2$ **L.** $y = 4 - x$

N. $2x - y = 5$ **P.** $y = 3^{-x}$ **R.** $y = 4 - x^2$

What do you call a difficult number plane sketch?

4–1–9–8–2   7–6–9–3–8   5–6–9–3–8

Part A General topics

Calculators are not allowed.

1 Round 4.082 35 to three significant figures.

2 Increase $800 by 25%.

3 Find the value of $4x^2$ if $x = -3$.

4 Write 0.000 562 in scientific notation.

5 Evaluate $10^4 \times 10^5 \div 10^0$

6 Find a simplified expression for the area of this rectangle.

$3x + 2$

$4x$

7 Write the formula for the circumference of a circle with radius r.

8 Find the median of these scores.

Stem	Leaf
0	2 3 3 8
1	0 4
2	1 9

9 Find the midpoint of the interval joining the points $(-2, 6)$ and $(8, 12)$ on the number plane.

10 Which quadrant does the point $(-3, -6)$ lie in?

Part B Graphing lines and curves

Calculators are allowed.

16–01 Graphing linear equations

11 Graph $y = 4 - 2x$ on the number plane and find its x-intercept and y-intercept.

16–02 Testing if a point lies on a line

12 Which of these points lie on the line $y = 3x - 8$? Select **A**, **B**, **C** or **D**.

   A $(-1, 11)$ **B** $(2, 2)$ **C** $(3, -1)$ **D** $(-1, -11)$

16–03 Horizontal and vertical lines

13 Graph each line on the same number plane.

   a $x = -2$ **b** $y = 5$ **c** $x = 3$ **d** $y = -2$

16–04 Finding the equation of a line

14 Find the equation of this table of values.

x	-1	0	1	2
y	-7	-4	-1	2

15 a What is the gradient of the line drawn?

   b What is the y-intercept of the line?

   c Write down the equation of the line.

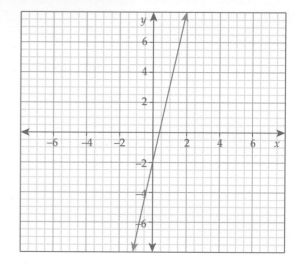

16–05 Graphing parabolas

16 Graph each quadratic equation on a number plane.

   a $y = x^2$ **b** $y = 2 - x^2$

16–06 Graphing exponential equations

17 Graph each exponential equation on a number plane.

   a $y = 2^x$ **b** $y = 4^{-x}$

16–07 Graphing circles

18 What is the equation of this circle?

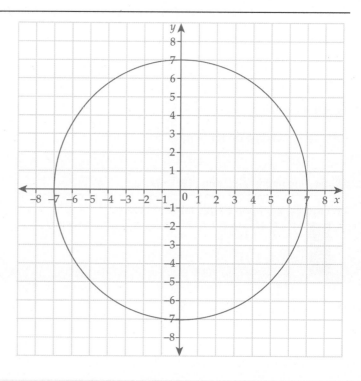

ISBN 9780170351058

CHAPTER 1

Exercise 1-01

1 B   **2** A

3 **a** 21   **b** 38   **c** 43   **d** 152
   e 40   **f** 59   **g** 62   **h** 97

4 C

5 **a** 91   **b** 91   **c** 127   **d** 175

6 **a** They are the same.
   b Just add 48 to part **c**'s answer.

7 **a** 15   **b** 26   **c** 41   **d** 56
   e 4   **f** 62   **g** 76   **h** 229

8 **a** 120   **b** 130   **c** 130   **d** 121
   e 237   **f** 942   **g** 651   **h** 592

9 **a** $75 + 11 = 75 + 10 + 1$
             $= 85 + 1$
             $= 86$
   b $278 - 41 = 278 - 40 - 1$
                $= 238 - 1$
                $= 237$

10 **a** 43   **b** 79   **c** 63   **d** 63
   e 91   **f** 205   **g** 123   **h** 194

11 **a** T   **b** F   **c** F   **d** T

Exercise 1-02

1 B   **2** D

3 **a** 24   **b** 40   **c** 21   **d** 16
   e 45   **f** 24   **g** 72   **h** 28
   i 48   **j** 40   **k** 42   **l** 48
   m 6   **n** 5   **o** 6   **p** 5
   q 7   **r** 7   **s** 5   **t** 7
   u 11   **v** 8   **w** 7   **x** 8

4 **a** 70   **b** 396   **c** 522   **d** 92
   e 4500   **f** 512   **g** 186   **h** 888
   i 320   **j** 430   **k** 2080   **l** 279
   m 280   **n** 168   **o** 275   **p** 288

5 **a** 29   **b** 34   **c** 21   **d** 131
   e 23   **f** 109   **g** 18   **h** 32
   i 160   **j** 49   **k** 190   **l** 38
   m 16   **n** 35   **o** 5.5   **p** 23.8
   q 25.4   **r** 326   **s** 980   **t** 0.065

6 **a** 290   **b** 1800   **c** 4600   **d** 6200
   e 180   **f** 21 000   **g** 180   **h** 1300

Exercise 1-03

1 C   **2** C

3 **a** 6   **b** −2   **c** 4   **d** 3
   e 0   **f** −2   **g** −5   **h** 4

4 **a** 3   **b** $7 + (-3)$ **c** $-7 + (-3)$

5 **a** 7   **b** 6   **c** −10   **d** −3
   e 10   **f** −11   **g** −8   **h** 14

6 **a** F   **b** T   **c** F   **d** F

7

×	−3	7	−5	9	−4
2	−6	14	−10	18	−8
−8	24	−56	40	−72	32
−6	18	−42	30	−54	24
11	−33	77	−55	99	−44

8 **a** −9   **b** −8   **c** 4   **d** −7
   e 9   **f** −5   **g** −6   **h** −11

9 **a** 9   **b** −27   **c** 81   **d** −100

10 22°C   **11** $3360

Exercise 1-04

1 C   **2** A

3 **a** ×   **b** ÷   **c** ÷
   d ÷   **e** ×   **f** ÷

4 **a** 38   **b** 57   **c** 79
   d 49   **e** −60   **f** 103

5 **a** −24   **b** −30   **c** 282
   d 116   **e** −14   **f** 30

6 **a** 9   **b** 7   **c** 208   **d** 234

7 **a** −132   **b** 33.6   **c** 22.2
   d 81.6   **e** −24   **f** 48.83
   g −64   **h** 17   **i** −4.6

8 $3820

Exercise 1-05

1 A   **2** C   **3** $\dfrac{9}{4}, \dfrac{11}{5}, \dfrac{9}{7}, \dfrac{7}{5}$

4 **a** $2\dfrac{1}{4}$ **b** $2\dfrac{1}{2}$ **c** $1\dfrac{4}{5}$ **d** $1\dfrac{7}{8}$
   e $1\dfrac{2}{3}$ **f** $1\dfrac{6}{7}$ **g** $2\dfrac{1}{6}$ **h** $1\dfrac{6}{9}$

5 **a** $\dfrac{4}{3}$ **b** $\dfrac{13}{4}$ **c** $\dfrac{23}{5}$ **d** $\dfrac{13}{8}$
   e $\dfrac{12}{5}$ **f** $\dfrac{23}{6}$ **g** $\dfrac{14}{9}$ **h** $\dfrac{14}{5}$

6 **a** 37   **b** 19

7 **a** 8   **b** 25   **c** 2   **d** 8

8 **a** T   **b** F   **c** T   **d** F

9 **a** $\dfrac{3}{10}$ **b** $\dfrac{1}{3}$ **c** $\dfrac{2}{3}$ **d** $\dfrac{4}{9}$
   e $\dfrac{5}{13}$ **f** $\dfrac{3}{5}$ **g** $\dfrac{7}{10}$ **h** $\dfrac{2}{5}$
   i $\dfrac{4}{7}$ **j** $\dfrac{1}{3}$

Exercise 1-06

1 B   **2** D

3 **a** T   **b** T   **c** T   **d** F
   e T   **f** F   **g** F   **h** T

4 $\dfrac{3}{24}, \dfrac{2}{3}, \dfrac{9}{12}, \dfrac{5}{6}$

5 $\dfrac{3}{4}, \dfrac{7}{10}, \dfrac{13}{20}, \dfrac{3}{5}$

6 **a** <   **b** >   **c** <   **d** >
   e >   **f** >   **g** <   **h** =

7 **a** Convert fractions to 8ths: $\frac{1}{2} = \frac{4}{8}, \frac{3}{4} = \frac{6}{8}, \frac{5}{8} = \frac{5}{8},$
$\frac{1}{4} = \frac{2}{8}, \frac{7}{8} = \frac{7}{8}$

b Convert fractions to 12ths: $\frac{1}{2} = \frac{6}{12}, \frac{3}{4} = \frac{9}{12}, \frac{2}{3} = \frac{8}{12},$
$\frac{5}{6} = \frac{10}{12}, \frac{5}{12} = \frac{5}{12}$

8 Brendon.

Exercise **1-07**

1 D   **2** C
3 **a** 5   **b** 9   **c** 5, 9.14
    d 8   **e** 3   **f** 8, 3, 5
4 **a** $\frac{3}{4}$ **b** $\frac{1}{2}$ **c** $\frac{14}{15}$ **d** $\frac{13}{24}$
    e $\frac{29}{40}$ **f** $\frac{35}{36}$ **g** $2\frac{4}{5}$ **h** $4\frac{1}{8}$
    i $6\frac{2}{5}$ **j** $2\frac{11}{12}$
5 **a** $\frac{1}{8}$ **b** $\frac{2}{15}$ **c** $\frac{5}{8}$ **d** $\frac{2}{9}$
    e $\frac{1}{3}$ **f** $\frac{1}{9}$ **g** $4\frac{1}{3}$ **h** $\frac{1}{9}$
    i $1\frac{1}{4}$ **j** $4\frac{1}{12}$
6 **a** $5\frac{7}{10}$ **b** $1\frac{4}{15}$ **c** $7\frac{31}{40}$
    d $4\frac{7}{8}$ **e** $2\frac{7}{12}$
7 **a** $\frac{19}{24}$ **b** $\frac{5}{24}$ **c** 2 pieces.

Exercise **1-08**

1 D   **2** A   **3** $\frac{2}{7} \times \frac{3}{5} = \frac{2 \times 3}{7 \times 5} = \frac{6}{35}$
4 **a** $\frac{1}{15}$ **b** $\frac{1}{4}$ **c** $\frac{3}{32}$ **d** $\frac{25}{96}$
    e $\frac{2}{3}$ **f** $\frac{5}{18}$ **g** 12   **h** $\frac{1}{9}$
    i $6\frac{2}{5}$ **j** $\frac{1}{24}$
5 $\frac{2}{3} \div \frac{3}{8} = \frac{2}{3} \times \frac{8}{3} = \frac{16}{9} = 1\frac{7}{9}$
6 **a** $1\frac{1}{2}$ **b** $\frac{1}{2}$ **c** $1\frac{1}{2}$ **d** $\frac{14}{27}$
    e $\frac{1}{4}$ **f** $\frac{3}{8}$ **g** $\frac{1}{27}$ **h** 1
    i $\frac{1}{25}$ **j** $\frac{1}{6}$
7 **a** $2\frac{4}{7}$ **b** $7\frac{13}{16}$ **c** $1\frac{11}{24}$
    d $3\frac{1}{14}$ **e** $8\frac{15}{16}$
8 **a** $\frac{1}{12}$ **b** $9500   **c** $2375

Language activity
FRACTION DISTRACTION

Practice test 1
1 450   **2** 45 000   **3** 1, 2, 3, 4, 6, 12
4 $3x + 2y$ **5** 9   **6** 10.8
7 $18   **8** $d = 14$ **9** $28xy$
10 12   **11** B
12 **a** 56   **b** 151   **c** 2173
13 B
14 **a** 48   **b** 19   **c** 908
15 **a** −3   **b** −11   **c** −5
    d 27   **e** −9   **f** −96
16 **a** 17   **b** 28   **c** −23
17 **a** $1\frac{3}{4}$ **b** $2\frac{1}{6}$ **c** $3\frac{7}{9}$
18 **a** T   **b** F   **c** F
19 **a** $1\frac{1}{12}$ **b** $\frac{7}{24}$ **c** $1\frac{3}{4}$
20 **a** $\frac{3}{10}$ **b** $3\frac{1}{3}$ **c** $\frac{2}{3}$
21 **a** 12   **b** $\frac{5}{24}$

CHAPTER 2

Exercise **2-01**

1 C   **2** B
3 **a** $\frac{1}{50}$ **b** $\frac{1}{20}$ **c** $\frac{13}{100}$
    d $\frac{1}{5}$ **e** $\frac{23}{100}$ **f** $\frac{3}{10}$
    g $\frac{19}{50}$ **h** $\frac{9}{20}$ **i** $\frac{31}{50}$
    j $\frac{7}{10}$ **k** $\frac{21}{25}$ **l** $\frac{19}{20}$
    m $1\frac{1}{20}$ **n** $1\frac{1}{5}$ **o** $1\frac{1}{2}$
4 **a** 0.02   **b** 0.05   **c** 0.13
    d 0.2   **e** 0.23   **f** 0.3
    g 0.38   **h** 0.45   **i** 0.62
    j 0.7   **k** 0.84   **l** 0.95
    m 1.05   **n** 1.2   **o** 1.5
5 **a** 75   **b** 100, 37.5
6 **a** 25%   **b** 40%   **c** 62.5%   **d** 30%
    e $41\frac{2}{3}$%   **f** 70%   **g** $55\frac{5}{9}$%   **h** $66\frac{2}{3}$%
    i 87.5%   **j** 65%
7 **a** 60%   **b** 2%   **c** 40%   **d** 18%
    e 32%   **f** 85%   **g** 12.5%   **h** 93%
    i 115%   **j** 250%
8 **a** 0.34, 35%, $\frac{3}{8}$ **b** $\frac{7}{10}$, 0.705, 74%
    c 25%, 0.255, $\frac{5}{18}$
9 **a** 76%, $\frac{3}{4}$, 0.745   **b** 34%, 0.315, $\frac{3}{10}$
    c 85%, 0.845, $\frac{5}{6}$

10

Fraction	$\frac{1}{20}$	$\frac{1}{10}$	$\frac{1}{5}$	$\frac{1}{4}$	$\frac{2}{5}$	$\frac{3}{5}$	$\frac{7}{10}$	$\frac{3}{4}$	$\frac{4}{5}$	1
Decimal	0.05	0.1	0.2	0.25	0.4	0.6	0.7	0.75	0.8	1
Percentage	5%	10%	20%	25%	40%	60%	70%	75%	80%	100%

Exercise 2-02

1 D   2 B
3 a $\frac{1}{4}$ b $\frac{1}{8}$ c $\frac{4}{5}$ d $\frac{1}{2}$

   e $\frac{1}{3}$ f 1   g $\frac{1}{10}$ h $\frac{2}{3}$

   i $\frac{3}{4}$ j $\frac{2}{5}$

4 a 4, 65   b 4, 160
5 a $80   b 2 yrs   c 32 km
   d 360 cm   e $1800   f 36 yrs
   g 400 L   h 2 days   i 64 m
   j $105   k $33   l 63 km
   m $60   n 60c   o $30
   p 30 c   q $28   r $2.80
6 a $17.50   b 1.92 m   c 51 m
   d 7 yrs   e 111.6 cm   f 34 yrs
   g 1.8 L   h 10.8 h   i 2.2 km
   j $617.40   k 21 h
   l $28 800
7 a 22 750   b 42 250
8 576 L   9 $1008

Exercise 2-03

1 D   2 B
3 a 15 out of 40 $= \frac{15}{40} = \frac{3}{8}$

   b 80 out of 220 $= \frac{80}{220} = \frac{4}{11}$

4 a $\frac{3}{5}$ b $\frac{3}{10}$ c $\frac{1}{4}$

   d $\frac{3}{5}$ e $\frac{9}{250}$ f $\frac{1}{16}$

   g $\frac{1}{4}$ h $\frac{1}{125}$ i $\frac{1}{6}$

5 a 60%   b 30%   c 25%
   d 60%   e 3.6%   f 6.25%
   g 25%   h 0.8%   i $16\frac{2}{3}$%
6 a 40, 45   b 50, 62.5
7 a 40%   b 37.5%   c 60%
   d 10%   e 6%   f $33\frac{1}{3}$%
   g 2.5%   h 37.5%   i 8%
8 a 20%   b 22°C   c 10%
9 Isabella, Jayden, Tahlia, Grace, Kurt, Blake

Exercise 2-04

1 D   2 B   3 add, subtract
4 a $216   b 1500 m   c 4.5 days
   d 5250 km   e $11 200   f 160 L
5 a $2000   b 72 mins   c 4500 km
   d 1800 kg   e $508.40   f 770 ha
6 a $493   b $637.50   c $1394
7 a $89.25   b $83.95   c $126
8 a $871.25   b $45 305
9 a $121   b $108.90, No
10 $14 400

Exercise 2-05

1 A   2 B
3 a T   b F   c F
4 a 25% of an amount = $450
     1% of the amount = $450 ÷ 25
               = $18
     100% of the amount = $18 × 100
                = $1800
   b 40% of an amount = $560
     1% of the amount = $560 ÷ 40
               = $14
     100% of the amount = $14 × 100
                = $1400
5 a $1800   b $600   c $1200
   d 6000 m   e 2000 L   f 210 m
   g 16 hrs   h $30 000   i 2000 kg
6 500 runs   7 $4800   8 2000
9 a $1120   b Find 26.5% of $1120.
10 a $150   b $82.50

Exercise 2-06

1 A   2 B
3

Cost price	Selling price	Profit or loss
$18.00	$26.00	$8.00 profit
$25.00	$38.00	$13.00 profit
$50.00	$36.00	$14.00 loss
$350.00	$290.00	$60.00 loss
$725.00	$760.00	$35.00 profit
$1290.60	$1182.10	$108.50 loss

4 a $400   b 50%   c $33\frac{1}{3}$%
5 a 60%   b 37.5%
6 a $500   b $66\frac{2}{3}$%   c 40%
7 a GST = $12, SP = $132
   b GST = $5.50, SP = $60.50
   c GST = $7.50, SP = $82.50
   d GST = $14, SP = $154

ANSWERS

e   GST = $3.20, SP = $35.20
f   GST = $2.60, SP = $28.60
g   GST = $4.80, SP = $52.80
h   GST = $3.60, SP = $39.60

8   a $92.40   b $42.35   c $57.75   d $107.80
    e $24.64   f $20.02   g $36.96   h $27.72

9   $412

Practice test 2

1   4   2   1, 2, 3, 6
3   9   4 $300
5   $\dfrac{2}{5}$ 6   8.8, 8.819, 8.89
7   $\dfrac{1}{6}$ 8   22 m
9   $-16xy^2$ 10 8
11 B   12 D   13 A
14 a   11 m   b   2 hours   c   52 L
15 a $\dfrac{4}{5}$ b $\dfrac{1}{3}$ c $\dfrac{3}{40}$
16 a   80%   b $33\dfrac{1}{3}$%   c   7.5%
17 a $660   b $2400
18 $767   19 $180
20 a $150   b   20%   21 $225.50

CHAPTER 3

Exercise 3-01

1   C   2   A
3   a $541.20   b $736.10
    c $700.91   d $942.31
4   a $1082.40   b $1472.20
    c $1401.82   d $1884.62
5   a   i $5729.17   ii $2644.23
    b   i $7410   ii $3420
    c   i $8873.33   ii $4095.38
    d   i $8045.83   ii $3713.46
6   a $43 087.20   b $51 304.80   c $56 322.50
7   a $755.77   b $1255.04
    c $1122.89   d $ 1372.36
8   a $4246.67   b $3685.28
    c $3302   d $4965.13
9   Holly   10 $59 507.20   11 $ 3150

Exercise 3-02

1   B   2   D
3   a $176.25   b $186.30
4   a $235   b $248.40
5   Normal wage = $29.40 × 35 = $ 1029
    Overtime = $29.40 × 1.5 × 3 + $29.40 × 2 × 2.5
                = $279.30
    Total wage = $1029 + $279.30
                = $ 1308.30
6   $916.50   7 $706.25
8   $1040.44   9 $1505.63

10

Day	Start	Finish	Normal hours	Time-and-a-half	Double-time
Monday	9.00 a.m.	4.00 p.m.	7	0	0
Tuesday	8.00 a.m.	3.00 p.m.	7	0	0
Wednesday	10.00 a.m.	7.30 p.m.	8	1.5	0
Thursday	8.30 a.m.	6.30 p.m.	8	2	0
Friday	8.15 a.m.	6.45 p.m.	8	2.5	0
Saturday	10.00 a.m.	3.30 p.m.	0	0	5.5
Sunday	10.50 a.m.	3.20 p.m.	0	0	4.5

Liam's wage is $1842.50.

ANSWERS

Exercise **3-03**

1   D   2   B   3 $15 650
4   a $1837.50   b $466.80   c $481.25   d $1408
5   a $840   b $1152   c $1073.52   d $871.68
6   a $5740   b $5356.80
    c $3050.91   d $4105.92
7   4 weeks wage = $865.50 × 4 = $3462
    Annual leave loading = 17.5% × $3462 = $605.85
    Total holiday pay = $3462 + $ 605.85 = $4067.85
8   a $546.42, $3668.82   b $511, $3431
    c $943.38, $6334.15
9   a $1299.04   b $8722.12
    c $5567.31

Exercise **3-04**

1   C   2   B
3   a $7144.80   b $11 576
    c $9132   d $10 002.40
4   a $6937.44   b $14 564.80
    c $44 508.16   d $19 662.51
5   a $10 023.25   b $34 034.20   c $18 808.70
    d $37 786   e   Nil   f $61 630
6   a $37 741.60   b $3813.02   c $3585.45
7   a $28147.50   b   20.9%

Exercise **3-05**

1   D   2   C
3   a $445   b $308   c $87   d $88
    e $1344   f $1026   g $342   h $338
4   $361.50
5   a $684   b $458.28

6

Name: Kate Keneally	Date: 16/06/16 to 20/6/16
Hourly rate: $27.80	Gross pay: $931.30
Hours worked: 33.5	Tax: $246.79
Tax rate: 26.5%	Net pay: $684.51

7   a $1940   b $440   c $1375
8   a $759.20   b   39.7%

Exercise **3-06**

1   B   2   C
3   a $200   b $2000
4   7, 3, 0.07, $105
5   a $127.50   b $1470
    c $572   d $14 580
6   a $150   b $1050
7   9, 0.0075, 7, 0.0075, 357
8   a $10.80   b $1008   c $992   d $2393.42
9   7 years.
10   5.4%

Exercise **3-07**

1   B   2   A
3   a $5250   b   60   c $40 800   d $46 050
4   a   i $863.20   ii $113.20
    b   i $1371.80   ii $121.80
    c   i $490   ii $71
    d   i $5273   ii $273
5   a $750   b   24   c $1980
    d $2730   e $530
6   $2550
7   a $110   b $702   c $812

Exercise **3-08**

1   C   2   B

3

Principal	Interest	Principal + interest
$3000	5% × $3000 = $150	$3000 + $150 = $3150
$3150	5% × $3150 = $157.50	$3150 + 157.50 = $3307.50
$3307.50	5% × 3307.50 = $165.38	$3307.50 + 165.38 = $3472.88
$3472.88	5% × 3472.88 = $173.64	$3472.88 + 173.64 = $3646.52

4

Year	Principal	Interest	Principal + interest
1	$6000	8% × $6000 = $480	$6000 + $480 = $6480
2	$6480	8% × $6480 = $518.40	$6480 + 518.40 = $6998.40
3	$6998.40	8% × 6998.40 = $559.87	$6998.40 + 559.87 = $7558.27

The amount $6000 grows to over 3 years at 8% p.a. compounded annually.

5 a $5304.50, $304.50   b $11 910.16, $1910.16
    c $13 167, $567   d $4840, $1840

6 $13 996.80

Exercise 3-09

1

Principal	Interest rate	Time	Final amount $A = P(1 + R)^n$	Compound interest $A - P$
$2000	8% p.a.	3 years	$2519.42	$519.42
$5800	7% p.a.	6 years	$8704.24	$2904.24
$9400	5% p.a.	3.5 years	$11 150.40	$1750.40
$12 700	4.5% p.a.	8 years	$18 060.68	$5360.68
$25 000	6% p.a. compounded monthly	8 months	$26 017.68	$1017.68

2 C   **3** A

4 a $4630.50, $630.50   b $6987.69, $1487.69
    c $15 751.04, $5151.04   d $24 252.02, $8052.02
    e $84 562.62, $20 562.62

5 a $429 187.07   b $446 496.98

6 Yes, he will have $5912.22.

Language activity

C	O	M	M	I	S	S	I	O	N			T			
O					U							E			
M				P	I	E	C	E	W	O	R	K			
P	A	Y			E			A				M			
O				G	R	O	S	S	G			S			P
U					A			E							R
N	E	T			N			D	E	P	O	S	I	T	
D				C	O	N	S	U	M	E	R		A		N
I				U				H		D			N		C
N				S	A	L	A	R	Y	I			I		I
T	I	M	E		T			E		L			N		P
E					I		E	A	R	N	I	N	G		A
R					O			R		D					L
E		B	O	N	U	S			R	A	T	E			
S									Y						
T	A	X	A	T	I	O	N								

Practice test 3

1 $4\frac{1}{2}$ **2** 4.15   **3** 7, 6, 3, 0, −2, −5

4 $6xy$ **5** 5   **6** $x = 116$

7 15:30   **8** $d = 18$ **9** $11 025

10 $75.60   **11** D   **12** B

13 C   **14** $986.40   **15** $19 150

16 a $476.42   b $3198.82

17 a $12 321   b $3826.54

18 a $528   b $2040

19 a $300   b $332.50

20 $37 680   **21** $57 058.31

22 a $86 393.26   b $18 393.26

CHAPTER 4

Exercise 4-01

1 C   **2** D

3 a $x - y$ b xy c $\dfrac{x}{y}$ d $x + y$

4 a T   b T   c F   d T

5 a $2a + 4$ b $16 - 3b$ c $2(m - n)$
    d $3(x + y)$ e $2n - 8$ f $m^2 - 3$
    g $3y + x$ h $2xy$ i $(m + n)^2$
    j $12 + g^2$ k $180 + 2d$ l $m^2 - 2n$
    m $15 - 3b$ n $h - 2j$

6 a $6a$ cents   b $4q   c $18d
    d $\dfrac{k}{5}   e $kw   f $\dfrac{d}{8}

7 a $4s$ b $b + 2c$ c $2(m + w)$

8 a s^2 b $\frac{1}{2}bh$ c mw

9 a $3w$ b $w - 3$ c $3w + 8$ d $w + 6$

10 a $4x$ b $2y$ c $4x + 2y$ d $x + y$

Exercise 4-02

1 B   **2** A

3 a F   b T   c F   d F

4 a 24   b −12   c −16   d 2
    e −40   f 32   g −28   h −48

5

	$4x - y$	$3x^2$	$\dfrac{4x}{y}$	$2(3x - y)$	$4xy^2$
$x = 2$, $y = 4$	4	12	2	4	128
$x = -1$, $y = 3$	−7	3	$-1\frac{1}{3}$	−12	−36
$x = 4$, $y = -2$	18	48	−8	28	64
$x = -1$, $y = -3$	−1	3	$1\frac{1}{3}$	0	−36
$x = 5$, $y = -2$	22	75	−10	34	80

6 a 8.4   b 56   c 25.16
    d 65.92   e 25   f 2

7 a $y = 2x + 1$

x	−1	0	1	2
y	−1	1	3	5

    b $y = 6 - x$

x	−1	0	1	2
y	7	6	5	4

c $y = 3x^2$

x	−1	0	2	3
y	3	0	12	27

d $y = \dfrac{3x}{2} - 2$

x	−2	0	2	4
y	−5	−2	1	4

8   377.0 cm³

Exercise 4-03

1   B   2   C
3   a $2x, 4x$ b $3a, a$ c $4w, -2w$
    d $7ab, 4ba$ e $3ab, 5ba$ f $-4x, 12x$
4   a   F   b   F   c   T   d   T
    e   F   f   T
5   a a b $8x$ c $9w - 3$
    d $10m$ e $6y$ f $2n + 9$
    g $-5a$ h $12m$ i $7t$
6   a $3m - 2n - 8m + n = 3m - 8m - 2n + n$
                         $= -5m - n$
    b $9x + 4y - 2y - x = 9x - x + 4y - 2y$
                         $= 8x + 2y$
7   a $9x - y$ b $5x - 4y$ c $13m - 3n$
    d $13m - n$ e $4x + 3y$ f $13a - 6b$
    g $23r - 9s$ h   7   i $xy - 7y + 9$
8   All the terms are unlike.
9   a $6x$ cents   b yr cents
10 a $10a + 2$ b $36m + 18$ c $6x + 10$

Exercise 4-04

1   B   2   D
3   a $24m$ b $-8xy$ c $-24mn$
    d $-36ab^2$ e $-18bc^2$ d f $24a^3$
    g $36m^4$ h $12uw^3$ i $32xy^3$
    j $-21a^2b^3$ k $8x$ l $-6b$
    m $5d$ n $-5ab$ o $-4bc$
    p $-3xy$ q $-6m$ r $\dfrac{12a}{c}$
    s $-\dfrac{3v}{x}$ t $-3r$ u $-\dfrac{5x}{y}$
4   a $25a^2$ b $21mn$ c $32xy$
5   a $2ab^2$ b $5x^2y$ c $-\dfrac{5m^2}{n^2}$
    d $12 - 12x$ e $6 + 2a$ f $8a^2$
6   $6n$ 7 $32x$

Exercise 4-05

1   C   2   C
3   a $2(3x - 4) = 2 \times 3x + 2 \times (-4)$
                  $= 6x + (-8)$
    b $-5(3a - 1) = -5 \times 3a + (-5) \times (-1)$
                   $= -15a + 5$

c $m(3m + 4) = m \times 3m + m \times 4$
              $= 3m^2 + 4m$
d $-x(2x - 4) = -x \times 2x + (-x) \times (-4)$
               $= -2x^2 + 4x$
4   a $4x + 8$ b $2a - 6$ c $10a - 5$
    d $18x + 6$ e $32 - 16a$ f $4a^2 - 3a$
    g $2x^2 + 3x$ h $2m^2 - 14m$ i $8w^2 + 24w$
    j $6a^2 - 10a$ k $-12a - 6$ l $-24 + 12m$
    m $-3x + 8$ n $-2a - 9$ o $-14w + 35$
5   a   both sides equal 28
    b   Teacher to check.
6   a $4(2y - 1) + 3y = 8y - 4 + 3y$
                       $= 11y - 4$
    b $16 + 2(3m - 1) = 16 + 6m - 2$
                       $= 6m + 14$
    c $3(3x - 2) - 2(x + 4) = 9x - 6 - 2x - 8$
                             $= 7x - 14$
    d $4(2a - 1) + 2(3a + 1) = 8a - 4 + 6a + 2$
                              $= 14a - 2$
7   a $6x - 26$ b $x - 1$ c $2a + 54$
    d $12x - 28$ e $9x - 15$ f $2a + 41$
    g $-14x - 4$ h $11x - 29$ i $-13a - 8$
8   $6w - 72v$

Exercise 4-06

1   D   2   C
3   a $2a$ b   4   c   6   d   9
    e $8m$ f $4a$ g $4w$ h $6m$
4   a $a + b$ b $x - 2y$ c $2m + 3n$ d $4b^2 - 3$
    e   3   f $12, y$ g $4, 4v$ h $3, 3n^2$
5   a $3(a + 4)$ b $9(p - 2)$ c $4(n + 7)$
    d $6(y - 5)$ e $9(a + 6)$ f $15(x - 3)$
    g $12(a + 6)$ h $16(c - 2)$ i $25(a + 3)$
    j $5(11x - y)$ k $a(a + m)$ l $b(a + c)$
    m $m(n + m)$ n $4(p - 6q)$ o $xy(z - 6)$
6   a $a + 3$ b $5 - 2y$ c $2m + 3q$
    d $14b, c$ e $14b, c$ f $15x, y$
7   a $3p(a - 5q)$ b $8s(2r + 3t)$
    c $5y(5 - 3y)$ d $16bc(2a - d)$
    e $4m(m + 12)$ f $11b(2b - 1)$
    g $19g(2f + g)$ h $3x(yz - 3)$
    i $13b(a^2 - 2b)$ j $9w(4w + v)$
    k $8rt(3r - 2)$ l $25mn(n + 2m)$
8   $16bc(4a^2 - b + 2a)$

Exercise 4-07

1   B   2   B
3   a   −3   b   −4   c   −5
    d   −5   e   −5   f $-8x$
    g $-14n$ h $-9v$
4   a $a + b$ b $x + 4y$ c $3m + n$ d $2b^2 + 1$
    e   −3   f $-7, y$ g $-9, 2v$ h $-3, 2n^2$

ANSWERS

5   a $-3(a + 6)$ b $-12(p + 2)$
    c $-5(n + 4)$ d $-6(y + 6)$
    e $-12(a + 7)$ f $-18(x - 2)$
    g $-21(a - 3)$ h $-6(c + 4)$
    i $-25(a - 3)$ j $-5(3x + 10y)$
    k $-a(a + b)$ l $-y(x - b)$
    m $-mn(1 - n)$ n $-9p(1 + 3q)$
    o $-xy(z - 8y)$

6   a $a + 3b$ b $x + 5y$ c $m - 2q$
    d $-16b, c$ e $-9b, 3c$ f $-25x, y$

7   a $-3p(a + 5q)$ b $-4s(2r + 5t)$
    c $-7y(x + 5y)$ d $-15bc(a - 2)$
    e $-8m(m - 8n)$ f $-11b(4b + a)$
    g $-3g(2f - 5g)$ h $-3x(yz + 4uv)$
    i $-7b(a^2 + 4bc)$ j $-8w(w - 3uv)$
    k $-9t(r^2 + 2v^2)$ l $-12mn(n - 4m)$

8   $-18bc(2a^2 + b - 4a)$

Language activity

EXPAND   ALGEBRAIC   EXPRESSION
LIKE   SIMPLIFY   SUBSTITUTE
UNLIKE   TERMS   FACTORISE
NEGATIVE   ADDING   PRONUMERAL
MULTIPLYING   COEFFICIENT   DIVIDING
FORMULA   SUBTRACTING

Practice test 4

1   $s^2 = p^2 + q^2$ 2 $A = \pi r^2$ 3   1.5%   4   26
5   3   6 $148   7   64   8   34 m
9   $-6a - 18$ 10 6   11 C   12 D
13 a -3.6 b   24.3   c   70.5
14 B
15 a $-11a + 4b$ b $18xy -4x - y$
    c $-4mn - m^2$
16 a $12xy$ b $-30a^3b$ c $63uv^2w$
17 a $-9a$ b $8x$ c $\dfrac{3b}{5a}$
18 $36mn$
19 a   F   b   T   c   F
20 a $9m - 5$ b $-16a^2 - 5a$
21 a $4(2a - 3b)$ b $4u(2u - 7v)$
    c $16ab(2b - a)$
22 a $-4(m + 5n)$ b $-2x(15 - 9y)$
    c $-25u(u + 3vw)$

CHAPTER 5

Exercise 5-01

1   D   2   C
3   a $a^2 = b^2 + c^2$
    b $n^2 = m^2 + o^2$
    c $e^2 = d^2 + f^2$
4   a   5   b   15   c   26
5   a $5^2 = 3^2 + 4^2$ b $15^2 = 9^2 + 12^2$
    c $26^2 = 10^2 + 24^2$

6   a $10^2 = 6^2 + 9^2$, No   b $25^2 = 7^2 + 24^2$, Yes
    c $22^2 = 12^2 + 16^2$, No
7   b

Exercise 5-02

1   C   2   A
3   $u^2 = 5^2 + 8^2$
    $u^2 = 89$
    $u = \sqrt{89}$
    ≈ 9.4 m
4   a   15 cm   b   26 m   c   25 m
    d $\sqrt{34}$ cm   e   30 m   f $\sqrt{53}$ m
5   d   5.8 cm   f   7.3 m
6   a   8.25 cm   b   20.50 m   c   19.78 m
7   7.2 m

Exercise 5-03

1   D
2   a   21 m   b   36 m   c   48 cm
    d $\sqrt{48}$ m   e   24 m   f $\sqrt{21}$ cm
3   a   18.76   b   11.49   c   8.01   d   13.92
4   6.61 m

Exercise 5-04

1   D   2   B
3   a   15 cm   b $\sqrt{287}$ m   c   30 m
    d   9 m   e $\sqrt{208}$ m   f $\sqrt{135}$ cm
4   a   11.10   b   14.33   c   23.24
5   a   4.2 m   b   10.2 m

Exercise 5-05

1   B   2   C
3   Does $40^2 = 24^2 + 32^2$?
        $1600 = 576 + 1024$
        $1600 = 1600$ Yes
    So the triangle is right-angled.
4   a   Yes   b   Yes   c   No   d   No
    e   Yes   f   Yes
5   a   5 and 12   b   14 and 48
    e   18 and 80   f   10 and 24
6   a

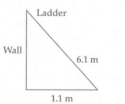

Ladder

Wall

6.1 m

1.1 m

    b   yes

ANSWERS

Exercise 5-06

1. B 2. B 3. D
4. 49, 625, is
5.
a	Yes	b	No	c	No	d	Yes
e	No	f	No	g	Yes	h	Yes
i	Yes	j	Yes	k	Yes	l	No

6. (sample answers)
   a. {6, 8, 10}   {9, 12, 15}
   b. {16, 30, 34}   {24, 45, 51}
   c. {18, 80, 82}   {27, 120, 123}
   d. {10, 24, 26}   {15, 36, 39}
7. a. $50^2 = 14^2 + 48^2$ b. {7, 24, 25}
8. a. T   b. F
9. Yes.

Exercise 5-07

1. C   2. 8.22 m   3. 7.12 m
4. a. 5 cm   b. 17 mm   5. 18 cm
6. 2.5 m   7. 12 m   8. 27 km

Language activity

Across
1. THEOREM   3. PYTHAGORAS
6. TRIANGLE   7. RIGHT
8. SIDES   9. ANGLE
10. SQUARE ROOT

Down
2. HYPOTENUSE   4. PYTHAGOREAN
5. TRIAD

Practice test 5

1. $r + 6$ 2. 0.058   3. $9y^2 - 18y$
4. 360   5. 2.21   6. 5
7. x-axis   8. 60 m³   9. 429
10. $10r + 3t$ or $2(5r + 3t)$
11. D   12. B
13. a. 17   b. $\sqrt{193}$
14. a. 6.24   b. 5.27
15. a. 13.9   b. 68.7   c. 50.0
16. a. Yes   b. No
17. a. Yes   b. No
18. 1.6 m   19. 100 cm

CHAPTER 6

Exercise 6-01

1. C   2. B
3. Hypotenuse = 17, Opposite = 8, Adjacent = 15
4. a.
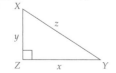

b.

c.

5. a. z b. q c. a
6. a. Hypotenuse = 5, Opposite = 4, Adjacent = 3
   b. Hypotenuse = 15, Opposite = 12, Adjacent = 9
   c. Hypotenuse = 25, Opposite = 24, Adjacent = 7
   d. Hypotenuse = b, Opposite = c, Adjacent = a
   e. Hypotenuse = r, Opposite = p, Adjacent = q
   f. Hypotenuse = r, Opposite = s, Adjacent = t
7. Teacher to check.

Exercise 6-02

1. A   2. C
3. a. $\sin\theta = \dfrac{\text{opposite}}{\text{hypotenuse}}$
   b. $\cos\theta = \dfrac{\text{adjacent}}{\text{hypotenuse}}$
   c. $\tan\theta = \dfrac{\text{opposite}}{\text{adjacent}}$
4. a. $\dfrac{4}{5}, \dfrac{3}{5}, \dfrac{4}{3}$ b. $\dfrac{3}{5}, \dfrac{4}{5}, \dfrac{3}{4}$
6. i. a. $\dfrac{p}{q}$ b. $\dfrac{r}{q}$ c. $\dfrac{p}{r}$
   ii. a. $\dfrac{c}{b}$ b. $\dfrac{a}{b}$ c. $\dfrac{c}{a}$
7. a. 25   b. 24   c. 7
   d. 24   e. 25   f. 24
8.

Exercise 6-03

1. C   2. D
3. $\tan 73° = \dfrac{m}{8}$
   $8\tan 73° = m$
   $m = 26.1668\ldots$
   $m \approx 26.17$
4. a. 7.25   b. 10.75   c. 9.38
5. a. 3.1   b. 14.2   c. 11.4
6. 8.01 m   7. 130.7 m

ANSWERS

Exercise 6-04

1 A   2 B
3 a T   b F   c F   d T
4 a 34°   b 57°   c 36°
5 $\angle P = 28°$
6 a 54°   b 61°   c 62°
7 70°   8 68°

Exercise 6-05

1 B   2 A
3 $\cos 42° = \dfrac{x}{19.4}$
$19.4 \cos 42° = x$
$x = 14.4170\ldots$
$x \approx 14.4$
4 a 9.23   b 5.78   c 7.42
   d 5.63   e 0.90
5 3.36 m   6 27.7 m   7 64.5 m

Exercise 6-06

1 A   2 C   3 $c = \dfrac{6}{\cos 53°}$
4 a 9.01   b 6.29   c 12.83
5 tan 42°, 3.1097…, 3.1
6 a 6.18   b 8.12   7 2.57

Exercise 6-07

1 C   2 B   3 12, 24.6243…°, 25°
4 a 34°   b 38°   c 52°
5 a 55°   b 78°   c 43°
6 $\angle Q = 50°$
7 38°   8 $\theta = 32°, \alpha = 58°$

Exercise 6-08

1 A
2 a elevation   b depression
   c depression
3 5.80 m   4 9.38 m   5 11.6 m   6 66°
7 No, it will only reach 3.53 m above the ground.
8 42°

Exercise 6-09

1 D   2 B
3 a 110°   b 317°   c 55°
   d 312°   e 253°   f 65°
4 a, b

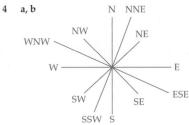

5 a

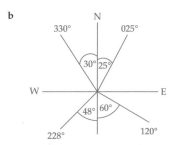

   b

6 a 000°   b 315°   c 090°   d 135°
   e 157.5°   f 022.5°   g 247.5°   h 337.5°

Exercise 6-10

1 C   2 D   3 3.5 km
4 a 31.8 km   b 31.8 km
5 a 6.13 km   b 310°
6 106.1 km   7 12.6 km   8 28.9 km

Practice test 6

1 $235.27   2 $\dfrac{9}{20}$ 3 20   4 24
5 straight angle   6 $36
7 $-6p^2 + 10p$ 8 No
9 8400   10 8   11 B   12 A
13 a 24.45   b 12.21
14 39°
15 a 11.6   b 44.2
16 a 16.5   b 10.5
17 a 67°   b 56°
18 63°
19 a 270°   b 045°
20 9.83 km

CHAPTER 7

Exercise 7-01

1	C		2	B			

3
a	15	b	18	c	9	d	90
e	30	f	9	g	3	h	7
i	6	j	9	k	15 : 24	l	15 : 25

4
a	2 : 3	b	1 : 6	c	4 : 3
d	3 : 1	e	10 : 3	f	5 : 12 : 15
g	3 : 4	h	4 : 3	i	3 : 2
j	12 : 25	k	15 : 8	l	10 : 9
m	1 : 8	n	1 : 9	o	1 : 10	p	5 : 2

5
a	1 : 4	b	1 : 4	c	10 : 1	d	1 : 4
e	4 : 25	f	1 : 25	g	1 : 4	h	4 : 3
i	6 : 5	j	7 :100	k	1 : 4	l	50 : 3
m	13 : 1250	n	1 : 6	o	3 : 4	p	5 : 8

6   $4 : 10 : 20 : 700 = 2 : 5 : 10 : 350$

Exercise 7-02

1	D		2	C	

3
| a | 1 : 2 | b | 3 : 1 | c | 2 : 3 |
4
| a | 14 : 9 | b | 2 : 1 | c | 1 : 2 |

3

Ratio	Total number of parts	Total amount	One part	Ratio
2 : 1	2 + 1 = 3	45	45 ÷ 3 = 15	30 : 15
3 : 1	3 + 1 = 4	48	48 ÷ 4 = 12	36 : 12
1 : 4	1 + 4 = 5	300	300 ÷ 5 = 60	60 : 240
2 : 5	2 + 5 = 7	560	560 ÷ 7 = 80	160 : 400
4 : 3	4 + 3 = 7	1470	1470 ÷ 7 = 210	840 : 630
5 : 3	5 + 3 = 8	24 000	24 000 ÷ 8 = 3000	15 000 : 9000

4 $135 : $90
5 $6120 : $8160
6 $2530
7 a $8200 b $4100
8 14
9 $39 000 : $65 000 : $52 000
10 $55 000, $77 000
11 $2930, $1172, $2344
12 Ashleigh 18, Sarah 15, Hayley 12

Exercise 7-05

1 D
2 A

---

5 $150, $400 6 168 7 12
8 $26 250
9 128 kg 10 104 11 125 12 $546
13 a 240 mL sugar, 360 mL milk b 350 mL
14 a 36 b $\dfrac{13}{23}$

Exercise 7-03

1	A		2	C

3
a	200 cm = 2 m
b	415 cm = 4. 15 m
c	250 cm = 2.5 m

4
| a | 25 cm | b | 75 cm | c | 6. 125 cm | d | 8.1 m |
5
| a | 3 cm | b | 4.6 cm |
6
| a | 1 : 400 | | | b | 1 : 300 |
| c | 3 : 400 000 | | | d | 1 : 30 |
7
| a | 8.3 m | b | 7.5 m |
8 9 m
9
| a | 2100 m | b | 2.1 km |

Exercise 7-04

1	B		2	A

3

Rate	Simplified rate
$4.50 for 9 apples	50 cents/apple
360 words in 8 minutes	45 words/minute
240 km in 3 hours	80 km/h
$1080 for 6 days at a hotel	$180/day
7 teachers for 168 students	24 students/teacher
5 wickets for 125 runs	25 runs/wicket
$18.60 for half an hour	$37.20/hour
1260 metres in 18 seconds	70 m/s

4 $62/h 5 7 km/L
6 a $165/day b $162/day c Nelson Hotel
7 65 beats/min 8 $23.50/h
9 96.1 km/h 10 639 words in 9 minutes.

Exercise 7-06

1 D   2 A

3 a $\dfrac{\$}{h}$ b \$644   c 16 hours

4 a 92 words/min   b $\dfrac{words}{minute}$

   c 45 mins   d 2760

5 12 L/100km

6 a $\dfrac{c}{\$}$ b \$8576   c \$54 500

7 a Marvellous Marg   b Jim's Jam

   c Mick's Marmite   d Hunger Honey

   e Cowper Coffee

8 Sam   9 West Wheels

10 a $\dfrac{\$}{kg}$ b \$5.88   c 1064 g

11 115 mins   12 11.9 L/100 km

Exercise 7-07

1 B   2 A

3

Distance	Time	Speed
150 km	3 h	50 km/h
750 km	5 h	150 km/h
1050 km	25 h	42 km/h
740 m	20 s	37 m/s
4950 m	90s	55 m/s

4 a 15 km   b 1.5 h   c 10 km/h

5 a 95 km/h   b 570 km   c 7.5 h

6 a 92 km   b 138 km/h   c 2300 m

   d 38.3 m/s

7 a 1.5 km   b 1500   c 25 m/s

8 a 240 km   b 240 km/h

   c 4000 m   d $66\dfrac{2}{3}$ m/s

Exercise 7-08

1 B   2 C

3 a 9 a.m.   b 32 km   c twice

   d 1.15 p.m.   e 5.3 km/h

   f 9.1 km/h on the way home, which is faster.

4 a 10 a.m.   b 22 km   c twice

   d 1.30 p.m.   e 4.7 km/h

   f 6.6 km/h on the way home, which is faster.

5 The group left at 9 a.m. and travelled 25 km in 1 hour. They stopped for half an hour. They then travelled another 25 km in 1 hour and reached their destination at 11.30 a.m. They stayed there for 1.5 hours and then travelled 50 km home in 2 hours, reaching home at 3 p.m.

Language activity

1 LIFETIME   4 : 4 = 1 : 1

2 SOMETHING   4 : 5

3 WITHOUT   4 : 3

4 EARTHQUAKE   5 : 5 = 1 : 1

5 BASKETBALL   6 : 4 = 3 : 2

6 UPSTREAM   2 : 6 = 1 : 3

7 FIREWORKS   4 : 5

8 PEPPERMINT   6 : 4 = 3 : 2

9 AIRPORT   3 : 4

10 SKATEBOARD   5 : 5 = 1 : 1

11 SUNFLOWER   3 : 6 = 1 : 2

12 SOUTHWEST   5 : 4

13 THUNDERSTORM   7 : 5

14 WIDESPREAD   4 : 6 = 2 : 3

15 AFTERNOON   5 : 4

16 HEADQUARTERS   4 : 8 = 1 : 2

Practice test 7

1 0.0325   2 $2b$ 3 108.5

4 $\dfrac{5}{16}$ 5 64   6 $6xy$

7 28°   8 \$1200   9 $\dfrac{2}{3}$ 10 36

11 D   12 B   13 A

14 a 4 : 6 : 1   b 4 : 11

15 a 11.1 m   b 78 cm

16 Ray \$12 000, Ed \$24 000, Pete \$36 000

17 a \$60/h   b 65c/apple

   c 78 words/min

18 94 beats/min

19 a \$6.15   b 720 g

20 a 1.25 km   b 1250 m

   c 20.8 m/s

21 distance

22 a 16 km   b 9:30 a.m.

   c 5.3 km/h

CHAPTER 8

Exercise 8-01

1 A   2 D

3 a isosceles   b equilateral   c scalene

4 a acute-angled

   b acute-angled

   c right-angled

5 a $x = y = 82°$ b $m = n = o = 60°$

   c $v = w = 45°$

6 a i c ii d, f

   b i t ii s, p

7 a $x = 79°$, angle sum of triangle

   b $n = 115°$, exterior angle of triangle

c $n = 98°$, angle sum of triangle
d $a = 24°$, angle sum of triangle
e $x = 107°$, exterior angle of triangle
f $n = 48°$, angle sum of triangle

8   a

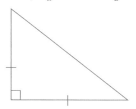

b

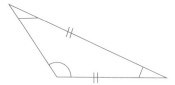

Exercise 8-02

1   B   2   B
3   a

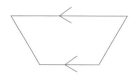

b

c

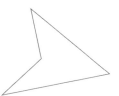

d

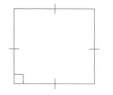

e

f

4   a   rectangle   b   kite   c   trapezium
5   a $x = 21$ b $p = 90$ c $n = 129$
    d $x = 118$ e $v = 124$ f $n = 92$
    g $n = 45$ h $n = 102$ i $x = 68$

Exercise 8-03

1   D   2   B
3   a – d

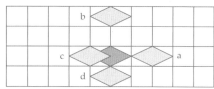

4

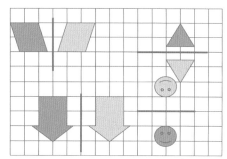

5   a – d

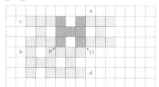

6 a

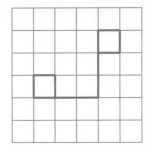

b

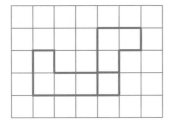

7 a

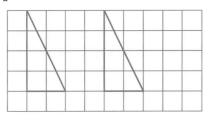

b

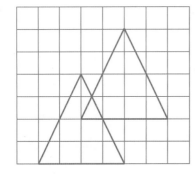

Exercise **8-04**

1 C   **2** C

3 *A* and *H*, translation; *B* and *E*, translation and reflection; *D* and *G*, rotation.

4 $AB = RP$, $BC = PQ$, $AC = RQ$,
$\angle A = \angle R$, $\angle B = \angle P$, $\angle C = \angle Q$

5 **a** $\angle J = \angle Q$, $\angle K = \angle R$, $\angle L = \angle S$, $\angle M = \angle T$, $\angle M = \angle P$
b *QRSTP*

6 Congruent shapes used: bird, fish
Transformations used: translation, reflection

Exercise **8-05**

1 B   **2** D

3 B and D (SSS); A and F (RHS); C and E (AAS)

4 **a**

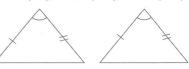

b

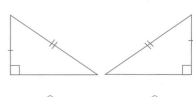

c

5 $\triangle ABC \equiv \triangle IHG$ (RHS) $\triangle RPQ \equiv \triangle DEF$ (SSS)
$\triangle WXY \equiv \triangle MON$ (SAS)

Exercise **8-06**

1 B   **2** D

3 **A, B, D**

4 A and M, C and N, D and G, F and H, J and I, K and Q, E and R.

5 $\triangle ABC \,|||\, \triangle XYZ$ and $\triangle DEF \,|||\, \triangle RPQ$

6 **a** Both 70°   **b** Yes   **c** Yes

7 **a** Both 60°   **b** Yes
c $DE = AB$, $EF = BC$, $DF = AC$
d $\triangle ACB$

Exercise **8-07**

1 C   **2** D

3 **a** 2   **b** $\dfrac{3}{2}$ **c** $\dfrac{5}{3}$ **d** $\dfrac{1}{2}$

e 3   **f** 3   **g** 5   **h** $\dfrac{3}{4}$

4 Teacher to inspect.

5 Teacher to inspect.

6 **a** A right-angled triangle with sides about the right-angle of 7.5 cm and 10 cm.

7 **b** A right-angled triangle with sides about the right-angle of 2 cm and $2\dfrac{2}{3}$ cm.

Exercise **8-08**

1 C

2 A

3

	Length	Height	Door length	Door height
Small house	4 units	3 units	1 unit	1.5 units
Big house	8 units	6 units	2 units	3 units
Scale factor	2	2	2	2

Yes, the houses are similar figures.

4   **a** $x = 9$ **b** $w = 6$ m

    c $y = 3$ **d** $x = 5$

5   5 cm

6   **a** PQR **b** $3^2 + 4^2 = 5^2$

    c $x = 10$

Language activity

1   SIMILAR   **2**   TRIANGLE

3   CONGRUENT

4   FIGURE   **5**   TRANSLATION

6   REDUCTION

7   QUADRILATERAL

8   ENLARGEMENT

9   REFLECTION

10   TRANSFORMATION

11   TEST

12   ROTATION

13   IMAGE

14   ANGLE

15   MATCHING

Practice test 8

1 $\dfrac{7}{15}$ **2** $2y(3xy + 1)$

3   4   **4** $16.55

5 $\angle QPR$ or $\angle RPQ$

6   acute   **7** $x^2 = 5^2 + 12^2$

8   13 cm   **9** $\dfrac{4a^2}{b}$

10 $10r + 6t$

11   C   **12**   B

13   **a**   parallelogram   **b**   parallelogram

14   **a** $x = 44$ **b** $m = 82$

15   rotation   **16** $SRUT$ **17**   RHS

18   **a**   A and D

19   **a**   3   **b** $\dfrac{3}{2}$

20 $x = 16$

CHAPTER 9

Exercise 9-01

1   C   **2**   D

3   **a**   38 cm   **b**   24.8 m   **c**   26.9 cm

4   21 cm, 52 mm, 16.2 cm, 28.8 m, 28 cm, 24.8 cm

5   **a**   36.4 cm   **b**   37.6 cm   **c**   22.4 cm

    d   48.4 cm   **e**   48.7 cm   **f**   35.6 cm

6   **a**   34 m   **b**   30 cm   **c**   29.6 m   **d**   85 m

7 a   10.8 m   **b** $345.60

Exercise 9-02

1   B   **2**   C

3   **a**   radius

    b   diameter

    c   arc

    d   chord

4   **a, b, c**

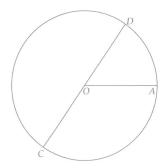

    d $OA = 4$ cm, $CD = 8$ cm

    e $CD = 2 \times OA$

    f   diameter = twice the radius.

5   **a**   diameter   **b**   segment   **c**   sector

    d   arc   **e**   radius   **f**   chord

    g   circumference   **h**   tangent

6   **a**   quadrant   **b**   tangent   **c**   radius

    d   arc   **e**   sector   **f**   chord

    g   segment   **h**   circumference

    i   semicircle   **j**   diameter

7   **a**   T   **b**   F   **c**   T

    d   F   **e**   T

8   **a**   diameter   **b**   arc   **c**   chord

    d   semicircle   **e**   tangent   **f**   sector

9   Teacher to check.

Exercise 9-03

1   C   **2**   B

3   **a**   18.0 m   **b**   14.3 cm   **c**   15.0 m

    d   23.9 m   **e**   43.2 m   **f**   61.4 mm

4   **a** $\left(\dfrac{7\pi}{2} + 7\right)$ m   **b** $(2\pi + 8)$ cm

5   **a**   33.1 m   **b**   17.3 cm   **c**   14.9 m   **d**   49.7 cm

6   **a**   19.5 m   **b** $819

Exercise 9-04

1   D   **2**   B

3   **a**   0350   **b**   1628   **c**   0536   **d**   2045

    e   0000   **f**   0925   **g**   0054   **h**   2355

4   **a**   8 a.m.   **b**   10:30 p.m.

    c   4:45 a.m.   **d**   8:18 p.m.

    e   3:20 a.m.   **f**   10:58 a.m.

    g   9:49 p.m.   **h**   12:46 a.m.

5

Time	Hours and minutes	Time	Minutes and seconds
5.2 hours	5 h 12 min	19.5 minutes	19 min 30 sec
192 minutes	3 h 12 min	225 seconds	3 min 45 sec
1246 minutes	20 h 46 min	386 seconds	6 min 26 sec
2.25 hours	2 h 15 min	54.4 seconds	0 min 54.4 sec

6   **a**   3 h   **b**   6 h   **c**   7 h   **d**   6 h
    e   16 min   **f**   8 min   **g**   10 min   **h**   23 min

7   **a**   4 h 25 min   **b**   1355

8   **a**   17 h 5 min   **b**   9 h 58 min   **c**   10 h 6 min
    d   12 h 3 min   **e**   8 h 19 min   **f**   18 h 16 min

9   6:02 p.m.

10   19 h 58 min

Exercise 9-05

1   B

2   D

3   **a**   07:23, 1 h 14 min   **b**   10:31, 35 min
    c   07:54, 1 h 5 min   **d**   09:43, 32 min
    e   12:52, 58 min   **f**   13:25, 48 min

4   **a**   Lapstone   **b**   Valley Heights
    c   Warrimoo   **d**   Linden

5   **a**   No   **b**   Yes   **c**   No   **d**   Yes

6   **a**   3, 7:52, 13:18 and 13:48
    b   Strathfield and Parramatta as all trains stop at these stations.

7   **a**   8:26 a.m.   **b**   8 h 52 min   **c**   8:13 p.m.

Exercise 9-06

1   D   **2**   D

3   **a**   5 p.m.   **b**   4 p.m.   **c**   3 a.m.   **d**   6 p.m.

4   **a**   4:30 a.m.   **b**   11:30 p.m. the day before
    c   10:30 p.m. the day before
    d   6:30 a.m.

5   3:20 a.m. Saturday

6   4 a.m. to 7:18 a.m.

7   **a**   5 p.m.   **b**   3 p.m   **c**   5 p.m.   **d**   5 p.m.

8   **a**   9:10 a.m.   **b**   8:40 a.m.
    c   9:10 a.m.   **d**   9:10 a.m.

9   2 p.m.   **10**   2 p.m.

Language activity

1   The average speed of the marathon runner.

2   The longer the race the more time he takes.

3   It's about time I finished.

Practice test 9

1 $240   **2** $5\frac{3}{4}$ **3** x^5

4   2   **5** $5m + 2$ **6**   isosceles

7   SAS   **8**   7 : 4   **9** $\frac{1}{81}$

---

10 $x = -5$ **11**   D   **12**   B

13   **a**   arc   **b**   quadrant

14   **a**   25.13 cm   **b**   15.08 m

15   36.85 cm

16   **a**   5 h 40 min   **b**   10 h 29 min

17   **a**   12 h   **b**   7 h

18   **a**   9:18   **b**   51 min

19   11:31

20   **a**   5:50 p.m.   **b**   3:55 p.m.

CHAPTER 10

Exercise 10-01

1   C   **2**   A

3   **a** $8 \times 8 \times 8$ **b**   8   **c**   3
    d   3, 11   **e**   9, 12   **f**   3, 9
    g   3, 5   **h**   8, 6   **i**   2, 4

4   **a** 9^7 **b** a^8 **c** n^7 **d** w^8
    e v^{11} **f** n^{13} **g** a^{15} **h** $6a^7$
    i $16n^7$ **j** $-32a^7$ **k** $-45n^7$ **l** $24w^8$
    m $-10v^9$ **n** $-21n^7$ **o** $30a^8$ **p** $-48a^7$

5   **a** 9^7 **b** a^8 **c** n **d** w^{10}
    e v^4 **f** n^4 **g** a^3 **h** $6a^8$
    i $3n^2$ **j** $4a^4$ **k** $3n^3$ **l** $-2w^{10}$
    m $-3v^8$ **n** $-7n^4$ **o** $7a^3$ **p** $-6a^7$

6   **a**   T   **b**   F   **c**   T   **d**   T

7   **a**   No   **b**   21952, 30.375

8   **a** $-24b^3c^6$ **b** $36a^7b^4$
    c $12m^7n^6$ **d** $21a^7c^5$
    e $-3b^6c^2$ **f** $-5a^8b^6$
    g $-4m^4n^2$ **h** $3a^5c^2$

Exercise 10-02

1   C   **2**   B

3   **a**   2, 6   **b**   5, x

4   **a** 2^{12} **b** 5^8 **c** 4^{12} **d** 3^{18}
    e 9^{40} **f** x^6 **g** n^{15} **h** m^{14}
    i w^{15} **j** a^{24} **k** 7^{15} **l** x^{12}
    m 5^{28} **n** q^{30} **o** p^{72}

5   **a**   F   **b**   T   **c**   T

6   **a** $9x^6$ **b** $8a^{12}$ **c** $16n^{12}$
    d $27m^{24}$ **e** $25c^{10}$ **f** $81w^{28}$
    g $343b^{12}$ **h** $64t^{10}$ **i** $-8a^{18}$
    j $125w^{18}$ **k** $9c^{16}$ **l** $-32q^{40}$

7   **a**   T   **b**   F   **c**   F

8   **a** $\dfrac{x^{12}}{8}$ **b** $\dfrac{m^6}{16}$ **c** $\dfrac{9}{w^{10}}$ **d** $\dfrac{9}{m^{16}}$
    e $\dfrac{x^{27}}{27}$ **f** $\dfrac{w^{15}}{-8}$ **g** $\dfrac{27}{c^{12}}$ **h** $\dfrac{-8}{n^{21}}$

9   **a**   F   **b**   T

10   **a** $\dfrac{8x^{18}}{27}$ **b** $\dfrac{m^{20}}{n^8}$ **c** $\dfrac{64}{27a^{12}}$ **d** $\dfrac{a^{36}}{b^{56}}$

Exercise 10-03

1 B 2 C

3 a T b F c F d T

4 a 1 b 1 c 1 d 1
  e 1 f 1 g 1 h 5
  i 1 j 7 k 1 l 8
  m 3 n 5 o 8 p 4
  q 2 r 4 s 6 t 6
  u 0 v 2 w 1 x 4

5 a 3 b 1, 4 c 2

6 a $\dfrac{1}{8}$ b $\dfrac{1}{9}$ c $\dfrac{1}{625}$ d $\dfrac{1}{64}$

  e $\dfrac{1}{n^5}$ f $\dfrac{1}{x^2}$ g $\dfrac{1}{m^4}$ h $\dfrac{5}{x}$

  i $\dfrac{1}{a^3}$ j $\dfrac{1}{25a^2}$ k $\dfrac{8}{a^6}$ l $\dfrac{1}{n^8}$

  m $\dfrac{1}{27b^3}$ n $\dfrac{2}{b^4}$ o $\dfrac{6}{x^5}$ p $\dfrac{1}{16n^2}$

  q $\dfrac{1}{27a^{12}}$ r $\dfrac{8}{b^3}$ s $\dfrac{1}{32c^5}$ t $\dfrac{6}{m^5}$

7 1 and 6, no

8 a $\dfrac{1}{49}$ b 1 c $\dfrac{4}{a^2}$ d $\dfrac{8}{b^5}$

Exercise 10-04

1 A 2 C

3 a F b F c F d T

4 a b^5 b n^7 c a^{11} d w^{12}
  e $15x^9$ f $12a^{13}$ g $45m^9$ h $24v^{14}$
  i n^6 j c k x^4 l v^6
  m $3a^6$ n $3m^3$ o $3c^6$ p $4w^8$

5 a 1 b 1 c 4 d −2
  e x^{20} f a^{35} g $243x^{15}$ h $8a^{18}$
  i 81 j 1 k 12 l 64

6 a n^5 b c^6 c x^{12} d 1
  e $15a^{10}$ f $8m^9$ g $8n^{15}$ h 8

7 a F b F c T d F
  e T f F g T h T
  i F j T

8 a $\dfrac{12}{a^5}$ b $\dfrac{2}{v^2}$ c 1 d $\dfrac{c^8}{256}$

  e $-18a^5b^5$ f $-6m^{12}n$ g $-189w^{15}$ h $-6x^6y^6$

Exercise 10-05

1 D 2 B

3 a 18 600 b 424 000 c 75 400
  d 1590 e 3 230 000 f 52 600
  g 48 500 000 h 328 000 i 208 000
  j 17 100 000 k 90 300 000 l 5 090 000

4 a 0.054 b 0.0029 c 0.74 d 8.5
  e 0.0021 f 0.084 g 0.000 83 h 0.50
  i 3.2 j 0.021 k 29 l 0.61

5 a 90 000 b 95 000 c 94 600

6 0.088

7 a 30 461 b 30 460

8 4.7

9 2 500 000

Exercise 10-06

1 A 2 A

3 a 4 b 5 c 5.6
  d 7.56 e 5.03, 8 f 4.2006, 8

4 a 4.5×10^5 b 7.26×10^{10} c 5.684×10^8
  d 6.08×10^7 e 2.98×10^8 f 3.4×10^6
  g 4.56×10^{10} h 9.85×10^9 i 6.35×10^6

5 a 54 000 b 816 000 c 980 000 000
  d 204 e 7 230 000 f 4618

6 a 36 000 b 6 950 000 c 800 000
  d 36 800 000 e 1 276 000 000
  f 420 000 000 g 1 120 000 000 000
  h 5 800 000 i 600 000 000 000

7 8×10^4, 6.8×10^6, 4.002×10^7, 7.48×10^7

8 2.6×10^9, 2.58×10^9, 3.42×10^6, 3×10^5

9 a 9.3824×10^4 b 2.58×10^5
  c 6.735×10^9 d 1.25×10^7

10 1×10^9 or 1.024×10^9 bytes.

Exercise 10-07

1 C 2 A

3 a −3 b −3 c 7.85
  d 4 e 1.5, −5 f 3.18, −6

4 a 5×10^{-4} b 6.2×10^{-4} c 7.08×10^{-5}
  d 7.004×10^{-3} e 4×10^{-7} f 9.2×10^{-7}
  g 1.86×10^{-9} h 8.6×10^{-6} i 2.3×10^{-3}

5 a 0.005 b 0.000 735 c 0.000 019
  d 0.000 000 28 e 0.804 f 0.000 042 25

6 a 0.021 b 0.000 062 8 c 0.0005
  d 0.000 004 05 e 0.000 000 001 115
  f 0.000 000 26 g 0.000 000 000 003 9
  h 0.000 006 12 i 0.000 000 07

7 3.06×10^{-7}, 3.6×10^{-7}, 4×10^{-3}

8 7×10^{-4}, 7.2×10^{-5}, 7.02×10^{-5}

9 a 2×10^{-1} b 3.1×10^{-3}
  c 7.4×10^{-6} d 2.5×10^{-5}

10 1×10^{-6}

Exercise 10-08

1 B 2 C

3 a 29 000 b 6 280 000 c 710 000
  d 0.000 042 5 e 0.000 0000 327 6
  f 0.000 86 g 9 120 000 000 000
  h 0.000 000 508 i 0.000 000 005 22

4 a 1.49×10^{13} b 6.24×10^{15}
  c 6.53×10^5 d 7.42×10^5
  e 7.60×10^8 f 4.21×10^{-5}
  g 4.04×10^6 h 2.14×10^7

5 a 4.5×10^4 b 1.5×10^9 c 2.0×10^{15}
  d 1.4×10^{22} e 1.4×10^{20} f 4.4×10^{-12}
  g 2.5×10^{14} h 1.6×10^{-15}

6 a 4.86×10^{16} b 5.40×10^{26}
   c 2.86×10^{6} d 3.03×10^{3}
7 1.44×10^{3} km
8 3.629×10^{-5}

Practice test 10

1 0.085   2 $6a(4b - a)$
3 right-angled (or scalene)
4 16 m^2 5 8 hours
6

7 1 hour 40 min   8 7:50 p.m.
9 $A = \pi r^2$ 10 $\dfrac{4}{11}$
11 C   12 B
13 a 3^{20} b n^{21} c $32x^{15}$
14 a $\dfrac{m^8}{n^4}$ b $\dfrac{a^8}{81}$ c $\dfrac{4096}{a^6}$
15 a 1   b 1   c 4
16 a $\dfrac{1}{25}$ b $\dfrac{2}{a^3}$ c $\dfrac{1}{3x^5}$
17 a $-12a^3b^3$ b $-3m^3n^3$ c $24a^8$
18 a 520 000   b 4 100 000   c 0.0033
19 a 4.522×10^{5} b 5.92×10^{6} c 2.3×10^{10}
20 a 1.45×10^{-2} b 7.24×10^{-4} c 1.2×10^{-8}
21 $5.6 \times 10^{-4}, 6 \times 10^{-4}, 7.2 \times 10^{6}$
22 a 901 800 or 9.018×10^{5} b 1.5×10^{20}

CHAPTER 11

Exercise 11-01

1 C   2 B
3 a 60 000   b 170 000   c 740
   d 2 900 000 e 4 100 000 f 1080
   g 40   h 0.1   i 36
   j 2   k 0.99   l 5.8
4 a 126 m^2 b 52.8 cm^2 c 20.25 m^2
   d 43.32 m^2 e 60.06 cm^2 f 218.4 cm^2
   g 42.56 cm^2 h 21.07 cm^2
5 a 108.36 cm^2 b 61.44 m^2 c 211.68 cm^2
   d 65.74 m^2
6 a 48.3 m^2 b 96 roses

Exercise 11-02

1 B   2 A
3 a 50.3 cm^2 b 24.6 m^2 c 160.6 cm^2
4 475 m^2
5 a 28.37 m^2 b 18.86 m^2 c 82.02 m^2
   d 68.50 m^2 e 23.13 m^2 f 96.37 m^2
   g 235.62 mm^2 h 43.98 m^2 i 6.41 cm^2
6 a 9.079 m^2 b 45

Exercise 11-03

1 C   2 A
3 a 221 cm^2 b 120 cm^2 c 120 cm^2
   d 76.96 m^2 e 116 m^2 f 75.1 cm^2
4 $86.11
5 a 6.15 m^2 b 20

Exercise 11-04

1 C   2 D
3 a

4 a 384 cm^2 b 116.98 cm^2
   c 215.86 m^2 d 500 m^2
5 a 105.84 mm^2 c 123.48 mm^2

Exercise 11-05

1 D   2 B
3 a

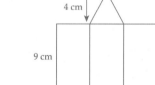

   b 168 cm^2
4 a 192 cm^2 b 252 m^2 c 150 cm^2
   d 600 m^2 e 43.44 m^2 f 920 m^2
5 192 cm^2

ANSWERS

Exercise 11-06
1 B
2 A
3 a

    b 653.5 cm²
4 a 402.1 m² b 34.3 m² c 280.6 cm²
  d 575.8 cm² e 276.5 m² f 713.5 m²
  g 85.2 m²
5 a 362.67 m² b 15 cans c $1027.50

Exercise 11-07
1 D   2 B
3 a 300 000 b 0.022 c 0.0086
  d 23.5 e 6.84 f 52 000 000
  g 7 800 000 h 76.4 i 44 700 000
  j 0.24 k 800 l 12 800 000
4 a 12 cm³ b 18 cm³ c 72 m³
  d 64 cm³ e 648 mm³ f 216 mm³
  g 12 cm³ h 92.5 m³
5 a 810 cm³ b 810 mL c 0.81 L

Exercise 11-08
1 C   2 B
3 a triangular prism

  b hexagonal prism

  c pentagonal prism

---

4 a A square and a rectangle
  b 60 m² c 10 d 600 m³
5 a 221 m³ b 726 m³
  c 225.6 m³ d 300 cm³
6 a 3801.3 cm³ b 1286.8 cm³ c 57.7 cm³
7 a 38.013 m³ b 38 013 L
8 a 296.1 m³ b 296 100 L c 11 844 fish

Language activity
1 KITE
2 RHOMBUS
3 PI
4 QUADRANT
5 SEMICIRCLE
6 TRAPEZIUM
7 DIAGONAL
8 AREA
9 PARALLELOGRAM
10 COMPOSITE
11 PRISM
12 SURFACE AREA
13 CYLINDER
14 VOLUME
15 CAPACITY

Practice test 11
1 0.079   2 y-axis
3 12   4 4.76×10^5
5 $y = 122$ 6 $800
7 $6h^8$ 8 26
9 14   10 21.5
11 A   12 C
13 B
14 a 10.6 cm² b 5.8 cm²
15 a 21.4 m² b 14.25 m²
16 182.56 cm²
17 a 13.5 cm² b 115.88 m²
18 276 cm²   19 213.2 m²
20 a 9 200 000 b 0.275 c 85 000
21 a 234 m³ b 450.3 cm³

CHAPTER 12

Exercise 12-01
1 B
2 a T b F c T
  d F e T f T
3 a census b sample c sample d sample
  e census f sample g census h census
4 a random b biased c random
  d biased e biased
5 a 750 b 18% c 9
6 Teacher to check.

Exercise 12-02

1　a　true　b　false　c　false
　　d　true　e　false　f　false
2　a　mode = 3, range = 6, median = 5, mean = 5.3
　　b　mode = 10 and 14, range = 20, median = 13, mean = 14.3
　　c　mode = 8, range = 4, median = 9, mean = 9.4
　　d　mode = 21, range = 17, median = 21, mean = 21
　　e　mode = 55, range = 18, median = 54.5, mean = 52.6
　　f　mode = 101, range = 5, median = 101, mean = 101.9
3　a　$21.29　b　$45　c　$17.33
　　d　It made it larger.
4　a　Reds: median = 46.05, range = 8.9
　　　Greens: median = 47.9, range = 11.2
　　　Blues: median = 52.65, range = 8.5
　　b　Reds　c　Greens
5　a　$890 000
　　b　mean = $891 068.75, median = $867 500
　　c　$1 450 800　d　The mean

Exercise 12-03

1　D　　　　2　D
3　horizontal　dot　score　hundreds　leaf
4　a

　b

　c

　d

　e

5　a　i　4, 8　　ii　6　　iii　5
　　b　i　11, 12　ii　12　iii　4
　　c　i　23　　ii　23　iii　9
　　d　i　54　　ii　54　iii　5

　e　i　16, 18　ii　16　iii　8
6　a　e shows a cluster　b　c has an outlier
7　a

Stem	Leaf
4	2　6　8
5	4　8
6	1　2　6　6　6
7	1　2　3　5　5
8	0　3　4

　b　66　　c　66　　d　65.7
8　a

Stem	Leaf
3	5　7　8　9
4	2　3　8
5	4　5　6
6	4　5　7　8
7	2　2　2　2　4　5
8	
9	7

　b　Range = 62, mode = 72
　c　$2667.86　　　　d　64
　e　Yes, in the 70s　f　Yes, 97

Exercise 12-04

1　D　　　　2　D
3

Score	Frequency	Cumulative frequency
30	3	3
31	6	9
32	8	17
33	10	27
34	12	39
35	4	43
36	2	45

Range = 6, mode = 34, median = 33.

4　a

Score	Frequency	Cumulative frequency	fx
21	2	2	42
22	7	9	154
23	1	10	23
24	2	12	48
25	1	13	25
26	1	14	26
27	1	15	27
28	2	17	56
29	3	20	87
30	4	24	120
Total	24		608

4　b　9　　c　22　　d　25.33　e　24.5

ANSWERS

5 **a**

Score	Frequency	Cumulative frequency
480	1	1
490	1	2
500	0	2
510	1	3
520	4	7
530	2	9
540	6	15
550	3	18
560	3	21
570	3	24
580	4	28
590	2	30
Total	30	

b $590   **c** $110   **d** $540   **e** $545

6

Score	Frequency	Cumulative frequency	fx
0	2	2	0
1	3	5	3
2	7	12	14
3	6	18	18
4	5	23	20
5	3	26	15
6	2	28	12
7	1	29	7
8	1	30	8
Total	30		97

6 **a** **i** 8   **ii** 2   **iii** 3   **iv** 3.23
   b $168

Exercise 12–05

1 D   **2** C
3 column joined equal line column lines
4 **a**

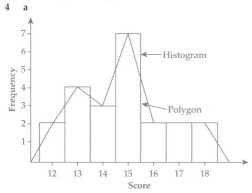

b Range = 6, mode = 15

c The range means the difference between the most number of messages sent to the least: 6. The mode means the most common number of messages sent: 15.
5 **a** Range = 5, mode = 12, mean ≈ 12.31
   b Range = 5, mode = 8, mean = 7.5

6 **a**

Score	Frequency
72	3
73	2
74	5
75	2
76	4
77	1
78	2
79	1

b

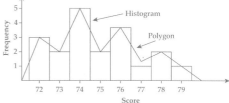

c Range = 7, mode = 74
d Mean = 74.9   **e** 50%

Exercise 12–06

1 B
2 **a** **i** negatively skewed
        ii cluster on the right, an outlier on the left
     b **i** symmetrical
        ii cluster in the centre, no outliers
     c **i** negatively skewed
        ii cluster at 140 and 150 stems, no outliers
     d **i** positively skewed
        ii cluster on the left, an outlier on the right
3 **a** positively skewed
     b mode = 13, range = 28
     c mean = 21.7
     d median = 19
     e in the tens
4 the mean, the other measures are not affected
   mode = 13, mean = 20.5, median = 19
5

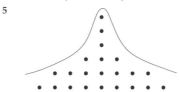

Exercise 12-07

1   D   2   C

3   **a**   Josie's range = 31, Ben's range = 22
     b   Ben
     c   Josie's mean = 25.5, Ben's mean = 36.1
     d   Ben

4

McDavids	Stem	Hungry Jim's
	0	28
9 5	29	8
7	30	6
8 7 4	31	
7 6	32	
	33	
5 2	34	2 5 8
	35	4 5 6 9
	36	6 7

   a   Hungry Jim's, 367   **b**   65
   c   McDavids = 317, Hungry Jim's = 354,
       Hungry Jim's
   d   McDavids = 315.45, Hungry Jim's = 345.09
   e   Hungry Jim's

5

Jess	Stem	Vicki
	6	9
8	7	4 8
9 6 4	8	3 6 8
9 5 3 2	9	3 4

   a   Jess = 21, Vicki = 25   **b**   Vicki, 69 s
   c   Jess = 90.5, Vicki = 84.5
   d   Vicki, yes
   e   Vicki, her times are generally lower, her
       median is lower and she showed the best
       time.

Exercise 12-08

1   B   2   D

3   **a**   13, 16, 3   **b**   6, 8, 2   **c**   21, 24, 3
     d   32.5, 36.5, 4   **e**   102.5, 107, 4.5

5   **a**   2   **b**   1.5   **c**   25   **d**   21

6   **a**   The Bluebells: 52,   The Red Ravens: 24;
       The Bluebells
    b   There is a large outlier for The Bluebells.
    c   The Bluebells: 11.5   The Red Ravens: 8.5
    d   The outlier is not involved in the calculation.
    e   The Red Ravens, as their interquartile range
       is lower and there is no outlier.

Exercise 12-09

1   D   2   B

3   The lowest score, the lower quartile, the median,
     the upper quartile and the highest score.

4   4, 5, 7, 8, 9

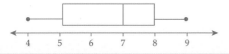

5   **a**

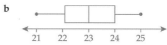

   b

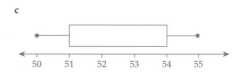

   c

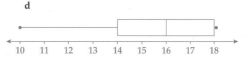

   d

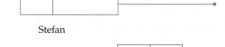

6   **a**   70   **b**   74, 64   **c**   10
     d   76   **e**   50%

7   **a**   Marisa

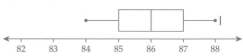

       Stefan

   b   Marisa: 6, Stefan: 4   **c**   Marisa: 3, Stefan: 2
   d   Stefan, further to the right
   e   Marisa: 83, Stefan: 86
   f   Stefan

Language activity

B	O	X	A	N	D	W	H	I	S	K	E	R		
A							I							
C					S	A	M	P	L	E				
K		H			T									
T		I		F	R	E	Q	U	E	N	C	Y		
O		S			I									
B		T		B	I	A	S		M					
A		O		U					O			R		
C		G		S	T	E	M	A	N	D	L	E	A	F
K		R		I		E			E			N		
		A		O		D	O	T				G		
		M	E	A	N		I					E		
							A							
					C	E	N	S	U	S				

ANSWERS

Practice test 12

1. 407
2. $\dfrac{7}{20}$
3. 68%, $\dfrac{2}{3}$, 0.65, 0.6
4. $2xy^2 + 14x^2y$
5. 5
6. $p = 70$
7. 10 m
8. $\dfrac{8}{11}$
9. 10.5 cm^2
10. 0
11. B
12. C
13. a 38 b 38 c 92
14. a 43.7 b Yes
15.

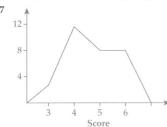

Stem	Leaf
3	0 5 5 5
4	0 0 0 5 5 5

16. a 4 b 3 c 4.5 d 4.625
17.

18. a Symmetrical b negatively skewed
19. Dr Watt: Range = 34, median = 53.5
    Dr Who: Range = 32, median = 46
    Dr Who has the shorter times.
20. Upper quartile = 47, lower quartile = 43,
    interquartile range = 4
21. a 16 b 10 c 75% d median

CHAPTER 13

Exercise 13-01

1. B
2. C
3. Many answers such as:
    a $x + 2 = 5$ b $2x + 1 = 7$ c $4x - 5 = 7$
4. a $3a - 5 = 19$ b $5x + 4 = 24$
    $3a - 5 + 5 = 19 + 5$ $5x + 4 - 4 = 24 - 4$
    $3a = 24$ $5x = 20$
    $\dfrac{3a}{3} = \dfrac{24}{3}$ $\dfrac{5x}{5} = \dfrac{20}{5}$
    $a = 8$ $x = 4$
5. a $a = 10$ b $w = 5$ c $n = 18$ d $x = 9$
   e $x = 3$ f $b = 6$ g $c = 9$ h $a = 7$
   i $x = 3\dfrac{1}{2}$ j $x = 4$
6. a $2\dfrac{1}{3}$ b $5\dfrac{1}{3}$ c 13 d 2
   e $-\dfrac{3}{4}$ f $-1\dfrac{1}{7}$ g 7 h $-2\dfrac{6}{11}$
   i 23 j -41 k -17 l -44
7. a T b F c T d F

Exercise 13-02

1. C
2. A
3. a $3x - 5$ b $4a + 9$ c $6 - 2x$
4. a $2x + 7$ b $2a - 5$ c $x + 18$
5. a $x = 12$ LHS = RHS = 31
   b $a = -7$ LHS = RHS = -19
   c $x = -4$ LHS = RHS = 14
6. a $3w + 18 = w - 4$
       $3w + 18 - w = w - 4 - w$
       $2w + 18 = -4$
       $2w + 18 - 18 = -4 - 18$
       $w = -11$
   b $5x - 19 = 4x + 17$
       $5x - 19 - 4x = 4x + 17 - 4x$
       $x - 19 = 17$
       $x - 19 + 19 = 17 + 19$
       $x = 36$
7. a 10 b 3 c -38
   d 9 e -12 f -1
   g 11 h -1 i -10
8. a T b F c T d $10\dfrac{1}{2}$
9. a $9\dfrac{1}{2}$ b $2\dfrac{1}{5}$ c -31
   d $2\dfrac{2}{5}$ e $5\dfrac{5}{6}$ f -6
   g $-\dfrac{2}{5}$ h $9\dfrac{1}{3}$ i -15

Exercise 13-03

1. B
2. C
3. a F b F c T d T
4. a $2(w - 4) = 14$
       $2w - 8 = 14$
       $2w - 8 + 8 = 14 + 8$
       $2w = 22$
       $w = 11$
   b $-3(a + 5) = 12$
       $-3a - 15 = 12$
       $-3a - 15 + 15 = 12 + 15$
       $-3a = 27$
       $a = -9$
5. a 15 b 4 c 11
   d 7 e 2 f 5
   g 2 h 2 i -3
6. a F b T c F d F
7. a $8\dfrac{3}{5}$ b $\dfrac{2}{3}$ c -2
   d $13\dfrac{1}{2}$ e $8\dfrac{3}{4}$ f $-2\dfrac{5}{14}$
   g $1\dfrac{1}{6}$ h $-\dfrac{4}{5}$ i $-3\dfrac{5}{6}$
8. $4(2x - 3) = 12$ and many other answers.
9. $3(2x + 1) = -15$ and many other answers.

Exercise **13-04**

1 B   **2** C

3 **a** T   **b** F   **c** T   **d** F

4 **a** ±7   **b** ±8   **c** ±12   **d** ±1

   e $\pm\sqrt{11}$ **f** $\pm\sqrt{19}$ **g** $\pm\sqrt{24}$ **h** $\pm\sqrt{39}$

   i No solution   **j** $\pm\sqrt{65}$

   k $\pm\sqrt{1000}$ **l** No solution

5 **e** ±3.3   **f** ±4.4   **g** ±4.9

   h ±6.2   **j** ±8.1   **k** ±31.6

6 **a** ±6   **b** ±3   **c** ±5   **d** ±1

   e $\pm\sqrt{\dfrac{7}{4}}$ **f** $\pm\sqrt{\dfrac{29}{2}}$ **g** $\pm\sqrt{53}$ **h** ±2

   i $\pm\sqrt{98}$ **j** No solution

   k $\pm\sqrt{300}$ **l** ±9

7 **e** ±1.32   **f** ±3.81   **g** ±7.28

   i ±9.90   **k** ±17.32

8 **a** T   **b** F   **c** F

   d T   **e** F   **f** F

9 **a** $n = \pm20.76$ **b** $w = \pm3.05$

   c $b = \pm2.32$ **d** $u = \pm31.62$

10 $x = \pm10\,000$

Exercise **13-05**

1 D   **2** B

3 **a** $n + 5 = 9, n = 4$ **b** $x - 7 = 12, x = 19$

   c $6w = 48, w = 8$ **d** $\dfrac{m}{4} = 13, m = 52$

4 7   **5** 13   **6** 5

7 14   **8** 6

9 15   **10** 17   **11** 19

12 5 hours

13 Katie is 36, Amy is 18.

14 Jasmine is 21, Trent is 7.

15 180 000

Exercise **13-06**

1 C   **2** B

3 **a**
$$3x + 1 < 13$$
$$3x + 1 - 1 < 13 - 1$$
$$3x < 12$$
$$\frac{3x}{3} < \frac{12}{3}$$
$$x < 4$$

   b
$$15 - 2x \geq 21$$
$$15 - 2x - 15 \geq 21 - 15$$
$$-2x \geq 6$$
$$\frac{-2x}{-2} \leq \frac{6}{-2}$$
$$x \leq -3$$

4 **a** $x < 8$

b $a > 3$

c $b \geq 9$

d $a \leq 8$

e $w > 5$

f $b > -3$

g $n < 16$

h $m > 4$

i $x \geq -9$

j $a < 1\dfrac{1}{2}$

k $x > -6$

l $b \leq -8$

5 **a** $x < 12$ **b** $v > -5$ **c** $x > 12$

   d $a > -4$ **e** $x \leq 8$ **f** $a > 10$

   g $n \leq -6$ **h** $b > -4\dfrac{1}{3}$ **i** $x \leq 5\dfrac{2}{3}$

6 Teacher to check.

Language activity

THEY ARE BOTH BALANCED

Practice test 13

1 $x = -6$ **2** 572.8   **3** 2238

4 17 hours

5 $x = 65$, corresponding angles on parallel lines

6 8   **7** 5   **8** 180 m²

9 $6\dfrac{2}{3}$ **10** 315°   **11** A   **12** C

13 **a** $a = -19$ **c** $n = -7$

14 **a** $a = -5$ **b** $x = 0$

15 Teacher to check.

16 **a** $x = \pm4$ **b** $n = \pm\sqrt{13}$

17 $y = \pm2.24$ **18** Jackson is 57, Amanda is 19.

19 **a** $x < 6$

   b $a \geq 3$

CHAPTER 14

Exercise **14-01**

1 A

2 **a** {Head, Tail}

   b {23, 25, 27, 29, 31, 33, 35}

   c {HH, HT, TH, TT}

3 **a** $\dfrac{1}{6}$ **b** $\dfrac{1}{6}$ **c** $\dfrac{1}{2}$ **d** $\dfrac{1}{2}$

   e $\dfrac{1}{2}$ **f** $\dfrac{1}{2}$ **g** $\dfrac{1}{2}$ **h** $\dfrac{2}{3}$

ANSWERS

4 a i $\dfrac{5}{11}$ ii $\dfrac{6}{11}$

   b 1

5 a Rolling 1, 2, 3, 5, 6

   b Not coming first but another a position.

   c Choosing a red sock.

d Not choosing a diamond, choosing heart, spade or clubs card.

f Not winning: either losing or having a draw.

6 a $\dfrac{1}{4}$ b $\dfrac{1}{3}$ c $\dfrac{5}{12}$

   d $\dfrac{3}{4}$ e $\dfrac{2}{3}$ f $\dfrac{7}{12}$

7 a $\dfrac{999}{1000}$ b 2000

8 a $\dfrac{1}{4}$ b $\dfrac{1}{4}$ c $\dfrac{3}{4}$

   d $\dfrac{1}{13}$ e $\dfrac{12}{13}$ f $\dfrac{1}{13}$

Exercise 14-02

1 $\dfrac{8}{20} = \dfrac{2}{5}$ 2 D

3 a

Score	Frequency	Relative frequency	Relative frequency (%)
1	28	$\dfrac{28}{180} = \dfrac{7}{45}$	15.6%
2	32	$\dfrac{32}{180} = \dfrac{8}{45}$	17.8%
3	27	$\dfrac{27}{180} = \dfrac{3}{20}$	15%
4	35	$\dfrac{35}{180} = \dfrac{7}{36}$	19.4%
5	31	$\dfrac{31}{180}$	17.2%
6	27	$\dfrac{27}{180} = \dfrac{3}{20}$	15%
Total	180	1	100%

b $\dfrac{7}{36}$ c 117   d $\dfrac{31}{180}$ e $\dfrac{149}{180}$

4 a, b Teacher to check.   c 1

5 a

Number of heads	Frequency	Relative frequency	Relative frequency (%)
0	9	$\dfrac{9}{80}$	11.25%
1	32	$\dfrac{32}{80} = \dfrac{2}{5}$	40%
2	28	$\dfrac{28}{80} = \dfrac{7}{20}$	35%
3	11	$\dfrac{11}{80}$	13.75%
Total	80	1	100%

b $\dfrac{39}{80}$ c 150

Exercise 14-03

1 B   2 A

3 a i 70 ii 26 iii 7 iv 39 v 5 vi 37

   b No

4 a i 57 ii 38 iii 15

     iv 0 v 4 vi 53

   b Yes

5

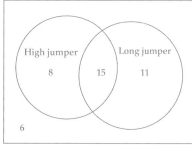

  a i 15 ii 8 iii 19

  b i $\dfrac{6}{40} = \dfrac{3}{20}$ ii $\dfrac{26}{40} = \dfrac{13}{20}$

6 a i 72 ii 34 iii 56

  b i $\dfrac{22}{72} = \dfrac{11}{36}$ ii $\dfrac{8}{72} = \dfrac{1}{9}$ iii $\dfrac{56}{72} = \dfrac{7}{9}$ iv $\dfrac{64}{72} = \dfrac{8}{9}$

7

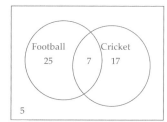

  a 54

  b i $\dfrac{17}{54}$ ii $\dfrac{7}{54}$ iii $\dfrac{7}{9}$ iv $\dfrac{16}{27}$

ANSWERS

Exercise 14-04

1 A   2 D

3

	Female	Male	Total
Blonde hair	12	7	19
Not blonde hair	26	19	45
Total	38	26	64

a 64   b 19   c 19

d $\dfrac{12}{64} = \dfrac{3}{16}$ e $\dfrac{26}{64} = \dfrac{13}{32}$

4

	Girls	Boys	Total
Dancing	18	14	32
Not dancing	6	8	14
Total	24	22	46

a 32   b 6   c $\dfrac{8}{46} = \dfrac{4}{23}$

d $\dfrac{18}{46} = \dfrac{9}{23}$

5 a

	TV	Not TV	Total
Movies	8	29	37
Not movies	21	2	23
Total	29	31	60

b 37   c 21   d $\dfrac{8}{60} = \dfrac{2}{15}$

e $\dfrac{50}{60} = \dfrac{5}{6}$

6

	Tennis	Not Tennis	Total
Soccer	36	22	58
Not soccer	18	14	32
Total	54	36	90

a 54   b 22

c $\dfrac{18}{90} = \dfrac{1}{5}$ d $\dfrac{36}{90} = \dfrac{2}{5}$

Exercise 14-05

1 A   2 B

3 a $\dfrac{5}{26}$ b $\dfrac{21}{26}$ c $\dfrac{3}{26}$

d $\dfrac{7}{26}$ e $\dfrac{4}{26} = \dfrac{2}{13}$

4 a $\dfrac{4}{30} = \dfrac{2}{15}$ b $\dfrac{6}{30} = \dfrac{1}{5}$

c $\dfrac{12}{30} = \dfrac{2}{5}$ d $\dfrac{22}{30} = \dfrac{11}{15}$

5 a i $\dfrac{48}{200} = \dfrac{6}{25}$ ii 0.24

b i $\dfrac{1}{4}$ ii 0.25

c Yes

d It should become closer to the theoretical probability.

6 a i $\dfrac{10}{26} = \dfrac{5}{13}$ ii $\dfrac{16}{26} = \dfrac{8}{13}$ iii $\dfrac{16}{26} = \dfrac{8}{13}$

b $\dfrac{7}{24}$

7 a i $\dfrac{16}{48} = \dfrac{1}{3}$ ii $\dfrac{24}{48} = \dfrac{1}{2}$

iii $\dfrac{32}{48} = \dfrac{2}{3}$ iv $\dfrac{32}{48} = \dfrac{2}{3}$

b $\dfrac{16}{46} = \dfrac{8}{23}$

Practice test 14

1 0.75   2 3   3 4.95×10^5

4 15 h 15 min

5 lengths of the rhombus' diagonals

6 $84   7 $a = 13$ 8 $45°$

9 8 and 10   10 4   11 B   12 D

13 a

Number of heads	Frequency	Relative frequency
0	6	$\dfrac{6}{40} = \dfrac{3}{20}$
1	26	$\dfrac{26}{40} = \dfrac{13}{20}$
2	8	$\dfrac{8}{40} = \dfrac{1}{5}$
Total	40	1

13 b $\dfrac{26}{40} = \dfrac{13}{20}$ c 65   d $\dfrac{6}{40} = \dfrac{3}{20}$

14

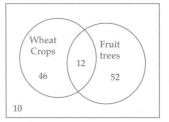

15 a 70   b 22   c 58

d $\dfrac{36}{70} = \dfrac{18}{35}$

16

	Smartphone	No smartphone	Total
Tablet	8	22	30
No tablet	36	4	40
Total	44	26	70

17 a

	Male	Female	Total
Passport	32	25	57
No passport	5	8	13
Total	37	33	70

b 57 **c** 5 **d** $\dfrac{25}{70} = \dfrac{5}{14}$

18 a $\dfrac{3}{10}$ **b** $\dfrac{7}{10}$ **c** $\dfrac{3}{10}$

19 a **i** $\dfrac{6}{32} = \dfrac{3}{16}$ **ii** $\dfrac{18}{32} = \dfrac{9}{16}$ **b** $\dfrac{12}{30} = \dfrac{2}{5}$

CHAPTER 15

Exercise 15-01

1 B

2 D

3 a T b F c F
  d T e F f T

4 a 11 b 17 c 8
  d 20 e 5 f 23

5 a $y = 3x$

x	−1	0	1	2
y	−3	0	3	6

b $y = x + 4$

x	−1	0	1	2
y	3	4	5	6

c $y = 2x − 1$

x	−1	0	3	4
y	−3	−1	5	7

d $y = 3x + 2$

x	−2	0	1	3
y	−4	2	5	11

e $y = 8 − x$

x	−2	0	2	4
y	10	8	6	4

f $y = 6 − 3x$

x	−3	0	3	6
y	15	6	−3	−12

g $d = 2c + 9$

c	−1	0	1	2
d	7	9	11	13

h $v = 4u − 7$

u	−1	0	1	2
v	−11	−7	−3	1

i $t = 14 − 2s$

s	−4	−2	0	2
t	22	18	14	10

j $q = 7p − 5$

p	−2	0	2	4
q	−19	−5	9	23

Exercise 15-02

1 D **2** B

3 a $y = x + 3$ b $y = \dfrac{x}{2}$ c $y = 5 − x$
  d $y = 3x$ e $y = x − 4$

4 a $y = x + 4$ b $y = x − 6$ c $y = 2x + 5$
  d $y = −2x + 3$ e $y = 3x + 1$ f $y = −2x + 2$

Exercise 15-03

1 3rd

2 B

3 a horizontal, y, right, down

4 a

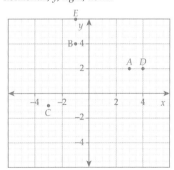

b A, D are in the 1st quadrant, B, E are in the 2nd quadrant and C is in the 3rd quadrant.

5 $A\,(2,3)$ $B\,(-2,1)$ $C(-2,-1)$ $D(2,-3)$ $E(4,1)$
     $F\,(0,3)$ $G\,(-3,5)$ $H\,(-1,-3)$ $I\,(2,0)$ $J\,(0,-2)$

6   **a** B and G **b** C and H
    c F and J **d** A and E

7   **a**   triangle   **b**   parallelogram
    c   square   **d**   hexagon

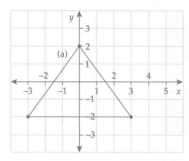

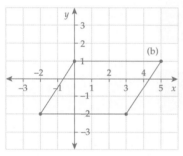

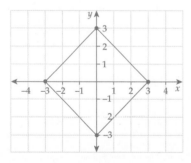

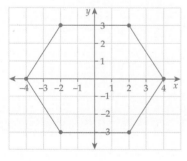

Exercise 15-04

1   C   **2**   A

3   **a**

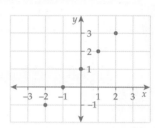

   b

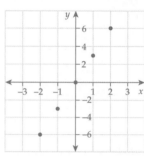

   c

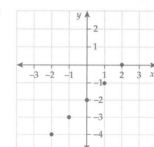

   d

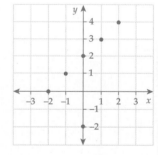

ISBN 9780170351058

e

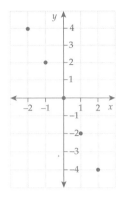

f

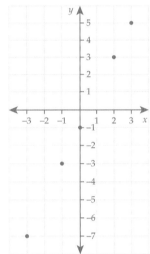

4 a They form a straight line. **b** Yes

5 a $y = x + 5$

x	−4	−2	0	1	2
y	1	3	5	6	7

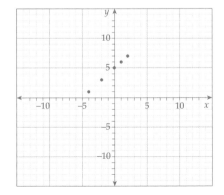

b $y = 3x - 2$

x	−2	−1	0	1	2
y	−8	−5	−2	1	4

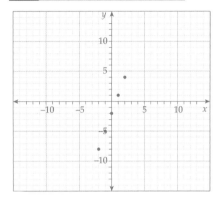

c $y = 5 - x$

x	−2	−1	0	1	2
y	7	6	5	4	3

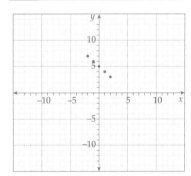

d $y = 4 - 2x$

x	−3	−1	0	2	3
y	10	6	4	0	−2

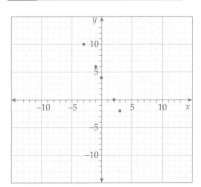

e $y = 4x - 10$

x	–1	0	1	2
y	–14	–10	–6	–2

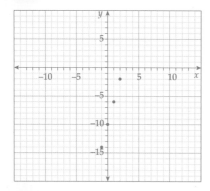

f $y = 5x - 3$

x	1	0	3	2
y	2	–3	12	7

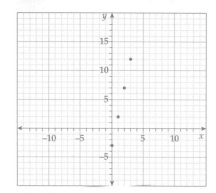

Exercise **15-05**
1 D   2 B
3 a F   b F   c T   d F
4 a $\sqrt{18}$ b $\sqrt{58}$ c $\sqrt{25} = 5$ d $\sqrt{29}$
5 a 8.5   b 5   c 13.5   d 13.4
   e 10.6   f 10   g 13.9
   h 12.6   i 8.6   j 13.9

Exercise **15-06**
1 B   2 C
3 a T   b F   c T   d T
4 a 7   b 4   c (7, 4)
5 a (2, 3)   b $(\frac{1}{2}, 2\frac{1}{2})$
6 (–2, 4)
7 a (2, 5)   b (4, 8)   c (4, 10)
   d $(5, 10\frac{1}{2})$ e (2, 7)   f (4, –6)

g $(\frac{1}{2}, 1)$ **h** (–2, 1)   **i** (–3, –1)
j (–1, –1)
8 a (6, 16)   b (7, –3)

Exercise **15-07**
1 C   2 B
3 a T   b F   c F   d T
4 a 7   b 7   c $\frac{7}{7} = 1$
5 a $\frac{4}{5}$ b $-\frac{3}{9} = -\frac{1}{3}$
   c $-\frac{6}{7}$ d $\frac{7}{6}$
6 a 1   b $-\frac{5}{8}$ c $-\frac{1}{7}$ d $\frac{7}{6}$
   e –4   f $\frac{5}{4}$ g $\frac{5}{3}$ h 1
   i $-\frac{3}{4}$ j $\frac{4}{7}$

Language activity
Across: 2 Midpoint   5 Quadrant
7 Horizontal   10 Gradient
Down: 1 Coordinates   3 Vertical
4 Length   6 Run
8 Rise   9 Table

Practice test 15
1 $\frac{3}{25}$ 2 $6a^2 - 24a$
3 –8   4 100 min
5 $7m + 20$ 6 1 030 000
7 12   8 91%
9 1000   10 $y = 80$
11 B
12 a $y = x - 3$

x	6	4	2
y	3	1	–1

   b $y = 2x + 3$

x	–1	0	1
y	1	3	5

13 $y = 2x + 1$
14 4th

15

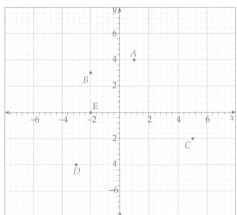

16

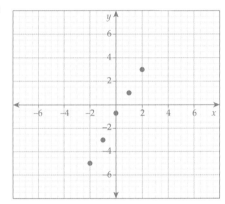

17 $y = 6 - 2x$

x	−1	0	1	2
y	8	6	4	2

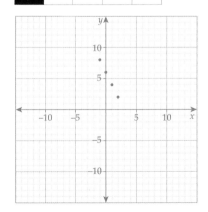

18 $\sqrt{20}$ units

19 (2, 3)

20 (1, −3)

21 2

22 −1

CHAPTER 16

Exercise **16–01**

1 C   2 D

3 a x-intercept = −4, y-intercept = 4

   b x-intercept = 2, y-intercept = −8

   c x-intercept = 2, y-intercept = 4

   d x-intercept = 2, y-intercept = 6

4 a $y = 3x$

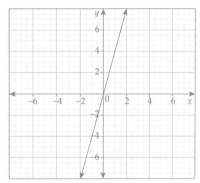

x-intercept = 0, y-intercept = 0

   b $y = x + 2$

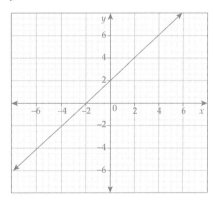

x-intercept = −2, y-intercept = 2

   c $y = 4 - x$

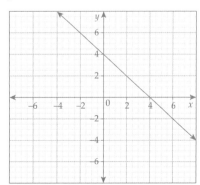

x-intercept = 4, y-intercept = 4

d $y = 4x - 1$

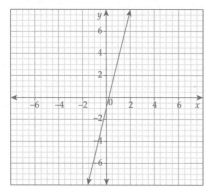

x-intercept $= \dfrac{1}{4}$, y-intercept $= -1$

e $y = 8 - 2x$

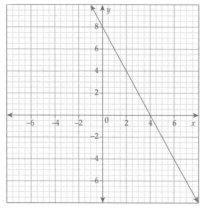

x-intercept $= 4$, y-intercept $= 8$

f $y = -2x + 4$

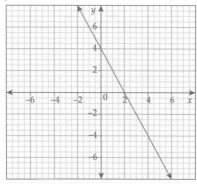

x-intercept $= 2$, y-intercept $= 4$

g $y = -2x + 6$

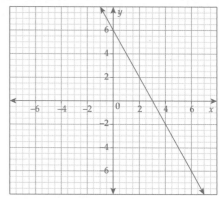

x-intercept $= 3$, y-intercept $= 6$

h $y = -2x - 1$

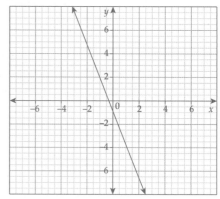

x-intercept $= -\dfrac{1}{2}$, y-intercept $= -1$

Exercise 16-02

1　C　　　　2　B

3　y-value, lies, false, lie

4　**b** and **e**

5

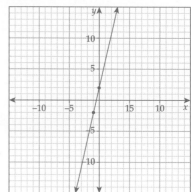

6　**a, c**

7 a No, yes **b** Yes, no **c** No, no **d** Yes, no
e Yes, no **f** No, yes **g** Yes, no **h** No, yes
8 (1, –1), (0, –4), (5, 11)

Exercise **16-03**

1 B **2** D
3 a F **b** T **c** T **d** F
4 a i $y = 4$ ii $y = 1$ iii $y = -2$
b i $x = -4$ ii $x = 2$ iii $x = 5$
5

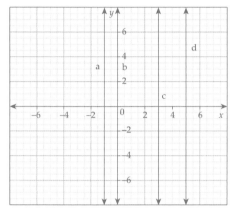

6

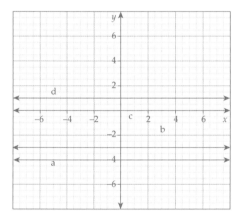

7 a $y = 4$ **b** $x = -6$ **c** $y = 4$ **d** $x = 5$
8 a (8, –4) **b** (–7, 6)

Exercise **16-04**

1 C **2** B
3 a $y = x + 9$ **b** $y = 3x$
c $y = 3x + 1$ **d** $y = 2x - 1$
4 a 4 **b** –4 **c** $y = 4x - 4$
5 a $y = 2x + 4$ **b** $y = -2x + 6$ **c** $y = -x + 4$
d $y = 4x + 2$ **e** $y = -6x + 4$ **f** $y = x + 6$

6 a

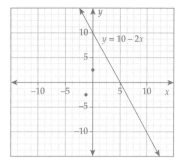

b gradient = –2, y-intercept = 10

Exercise **16-05**

1 D **2** B
3 a blue **b** red **c** pink **d** light blue

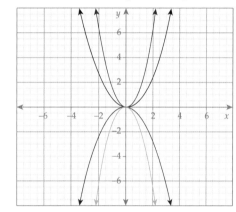

4 a and **b**

5 a

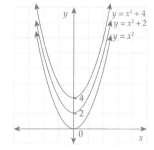

b

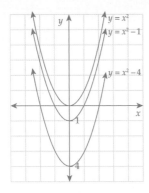

c

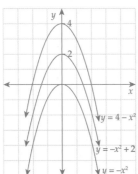

d

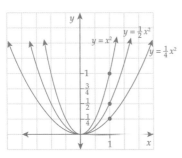

e

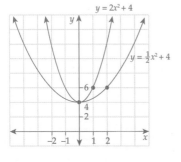

f

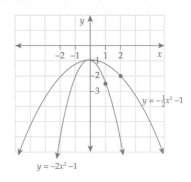

Exercise 16-06

1 D

2 a blue **b** red **c** pink **d** light blue

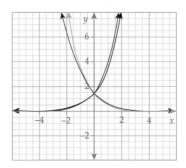

3 a and **b** are increasing and **c** and **d** are decreasing.

4 B

5 a T **b** T **c** T
  d T **e** F **f** T

6 a

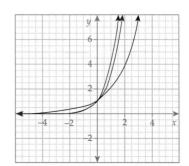

b

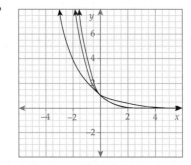

ISBN 9780170351058

c

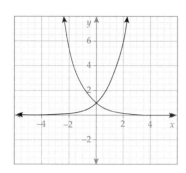

Note: $y = 3^{-x}$ and $y = \left(\dfrac{1}{3}\right)^x$ are the same graph.

d

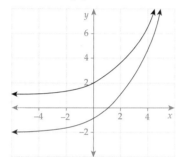

e

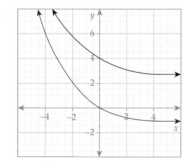

7   Asymptote for **c** is the x-axis.
Asymptote for **e** is the $y = 3$ and $y = -1$.

Exercise **16-08**

1   B

2   **a**

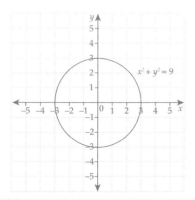

b

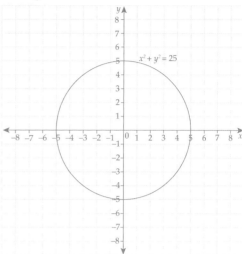

$x^2 + y^2 = 25$

c

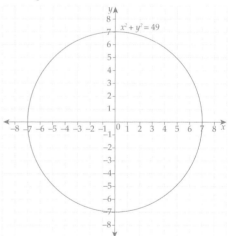

$x^2 + y^2 = 49$

d

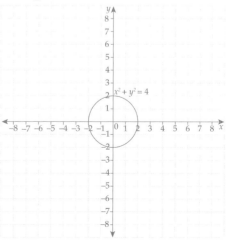

$x^2 + y^2 = 4$

e

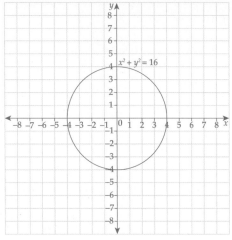

$x^2 + y^2 = 16$

f

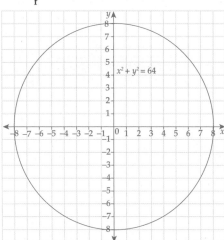

$x^2 + y^2 = 64$

g

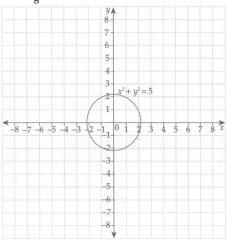

$x^2 + y^2 = 5$

h

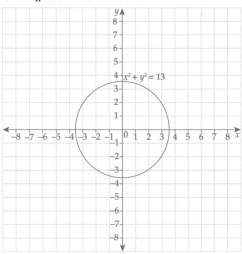

$x^2 + y^2 = 13$

i

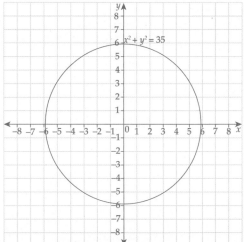

$x^2 + y^2 = 35$

3 A

4 **a** $x^2 + y^2 = 1$ **b** $x^2 + y^2 = 81$
    c $x^2 + y^2 = 36$ **d** $x^2 + y^2 = 9$
    e $x^2 + y^2 = 49$ **f** $x^2 + y^2 = 25$
    g $x^2 + y^2 = 121$ **h** $x^2 + y^2 = 64$
    i $x^2 + y^2 = 100$

5 **a** $x^2 + y^2 = 49$ **b** $x^2 + y^2 = 16$

Language activity
PLANE BRAIN DRAIN

Practice test 16
1 4.08
2 $1000
3 36
4 5.62×10^{-4}
5 1 000 000 000 or 10^9
6 $12x^2 + 8x$

7 $C = 2\pi r$

8 9

9 (3, 9)

10 3rd quadrant

11

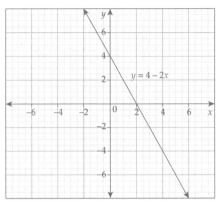

x-intercept 2, y-intercept 4

12 D

13

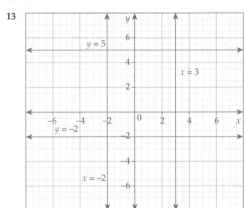

14 $y = 3x - 4$

15 **a** 4.5 **b** −2 **c** $y = 4.5x - 2$

16 **a**

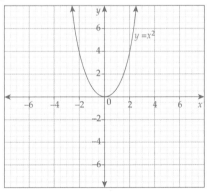

b

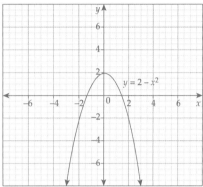

17 **a**

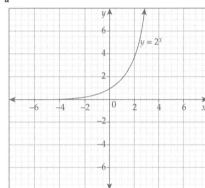

b

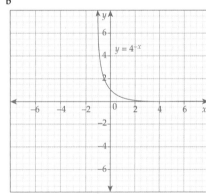

18 $x^2 + y^2 = 49$